Progress in Nonlinear Differential Equations and Their Applications
Volume 17

Serge Alinhac

Blowup for Nonlinear Hyperbolic Equations

Birkhäuser
Boston • Basel • Berlin

Serge Alinhac
Département de Mathématiques
Université de Paris-Sud
91405 Orsay Cedex
France

Library of Congress Cataloging In-Publication Data

Alinhac, S. (Serge)
    Blowup for nonlinear hyperbolic equations / Serge Alinhac.
      p.   cm. -- (Progress in nonlinear differential equations and
  their applications : v. 17)
    Includes bibliographical references (p. -) and index.
    ISBN 0-8176-3810-5 (acid-free). -- ISBN 3-7643-3810-5 (acid-free)
    1. Differential equations, Hyperbolic--Numberical solutions.
  2. Cauchy problem. 3. Blowing up (Mathematics) I. Title.
  II. Series.
QA377.A557   1995   95-2743
515'.353--dc20   CIP

Printed on acid-free paper
© 1995 Birkhäuser Boston

Birkhäuser

ISBN 0-8176-3810-5
ISBN 3-7643-3810-5
Typeset from disk by TeXniques, Brighton, MA
Printed and bound by Quinn-Woodbine, Woodbine, NJ
Printed in the U.S.A.

9 8 7 6 5 4 3 2 1

CONTENTS

CHAPTER III. Semilinear Wave Equations

CHAPTER IV. Quasilinear Systems in One
                     Space Dimension

CHAPTER V. Nonlinear Geometrical Optics
                     and Applications

Foreword

The content of this book corresponds to a one-semester course taught at the University of Paris-Sud (Orsay) in the spring 1994. It is accessible to students or researchers with a basic elementary knowledge of Partial Differential Equations, especially of hyperbolic PDE (Cauchy problem, wave operator, energy inequality, finite speed of propagation, symmetric systems, etc.).

This course is not some final encyclopedic reference gathering all available results. We tried instead to provide a **short synthetic view** of what we believe are the main results obtained so far, with self-contained proofs. In fact, many of the most important questions in the field are still completely open, and we hope that this monograph will give young mathematicians the desire to perform further research.

The bibliography, restricted to papers where blowup is explicitly discussed, is the only part we tried to make as complete as possible (despite the new preprints circulating everyday); the references are generally not mentioned in the text, but in the Notes at the end of each chapter.

Basic references corresponding best to the content of these Notes are the books by Courant and Friedrichs [CFr], Hörmander [Hö1] and [Hö2], Majda [Ma] and Smoller [Sm], and the survey papers by John [Jo6], Strauss [St] and Zuily [Zu].

Finally, I would like to thank all my colleagues and students who helped me improve this text, especially C. Zuily.

INTRODUCTION

This book deals with the phenomenon of blowup of classical solutions of global Cauchy problems for hyperbolic equations or systems.

1. We will consider either quasilinear systems of the form

$$(1) \qquad \partial_t u + \sum_{j=1}^{n} A_j(u)\partial_j u + B(u) = 0$$

or quasilinear wave equations of the form

$$(2) \qquad \partial_t^2 u + \sum_{0 \le i,j \le n} g_{ij}(\nabla u)\partial_{ij}^2 u + F(u, \nabla u) = 0.$$

The systems are assumed to be symmetric (or else at least strictly hyperbolic), and the principal parts of the equations are close to the wave operator

$$\partial_t^2 - \Delta_x.$$

By "global" Cauchy problem we mean that the given initial values are smooth with compact support. This seems to be the easiest case to consider because decay and decoupling tend to make the solution simpler with time. The case of periodic data or problems with boundaries is much more intricate and will not be considered here.

Our attention will not be restricted to the case of one space dimension ($n = 1$), though of course more results are available in this case. On the contrary, we will try to emphasize multidimensional results which appear to us as the first contours of a foggy landscape still to be described.

2. The striking feature of nonlinear global Cauchy problems with smooth data is that in general the solution will not remain smooth for all time (thus, "global" refers to space variables only). This phenomenon is called "blowup" for convenience, though it is generally hard to say what exactly "blows up".

The first concept which can be defined in this context is the "lifespan" $\bar{T}$ of the solution, that is, the biggest time up to which the smooth solution exists. Even the simplest question whether $\bar{T} < \infty$ or $\bar{T} = \infty$ is not yet properly understood for multidimensional problems (that is, $n \ge 2$).

If $\bar{T} < \infty$, it remains to say where the solution stops being smooth, and what is the exact mechanism of the blowup; one can also try to continue the solution as a weak (that is, nonclassical) solution after $\bar{T}$. We will not discuss this later aspect.

Even when we know $\bar{T} = \infty$, the qualitative behavior of the solution for large times is often poorly understood.

3. Our approach is based on the display and study of two **local blowup mechanisms**, which we call the "Ordinary Differential Equation mechanism" and the "Geometric Blowup mechanism".

The simplest occurrence of the first is the blowup of the solution of a system of Ordinary Differential Equations, such as $y' = y^2, y(0) > 0$, for example. The classical example of the second is provided by scalar quasi-linear equations, which can be handled by the method of characteristics. Thus the first chapter of this book is devoted to explaining what these mechanisms are.

The first mechanism is characterized by a nonlinear self-increase of the solution in an influence domain leading to the blowup locus (where the solution itself becomes infinite). Besides the systems of ODE, we study semilinear hyperbolic systems in one space variable and wave equations in higher dimensions. By calling this mechanism the "ODE mechanism", we do not mean that some ODE are responsible for the blowup, but that it is analogous to what happens for ODE, intervals being replaced more generally by influence domains.

The "Geometric Blowup mechanism" is defined in the very general context of quasi-linear systems in any number of dimensions. It is a generalization of the "focusing of rays" known for scalar equations. Its main feature is that the solution remains continuous at the blowup locus, only its gradient becoming infinite there.

We end this first chapter with a brief discussion of occurrences where these two mechanisms are combined, or compete.

The remaining chapters deal with **global** situations, for which we try to show that the solution behaves, locally near blowup points, according to the previously displayed **local** models.

4. Before distinguishing the various cases, we discuss in Chapter II the concept of lifespan and the striking fact that a smooth solution does not become gradually nonclassical, but undergoes a **brutal change** of regularity at time $\bar{T}$ (the lifespan). This brutal change is usually described as the "blowup criterion" in the literature. This result can be viewed as a nonlinear propagation of regularity (from the past to the future), and is obtained using energy methods.

After that, we address the question: "blowup or not ?" We give typical examples (Burgers' equation, semilinear and quasilinear wave equations, Euler system) and a very rough classification of the many methods used to give positive answers to this question. We distinguish the "functional methods" and the "comparison and averaging methods". We will see that in general these methods hardly provide more than just a "yes", giving in particular no information on the blowup mechanism at time $\bar{T}$.

5. In Chapter III, we study specifically semilinear wave equations (mostly the case where $F(u, \nabla u) = F(u)$, for rather special F).

In this case, a more refined concept than the lifespan can be introduced: the **maximal influence domain**. The shape of this domain and the behavior of the solution when approaching its boundary is a rich field to study. Of course, this boundary is space-like or characteristic, and one expects the solution on it to become infinite.

In the most favorable cases, the presently available results give the picture of a blowup taking place uniformly (that is, everywhere and with a constant rate) on a differentiable space-like surface, very close to the description by the ODE mechanism discussed in Chapter I. We close the chapter by just quoting a typical example of a sharp estimate of the lifespan.

6. Chapter IV is devoted to quasilinear systems or equations in one space dimension.

For 2×2 systems, assuming that Riemann invariants can be used to diagonalize the system, finite time blowup is established unless all the eigenvalues are linearly degenerate. If blowup occurs after the decoupling of the two modes, it is just a well understood scalar phenomenon; if it occurs before, we show that it is of a geometrical nature in the sense of Chapter I.

For general systems with genuinely nonlinear eigenvalues, no diagonalization is available, and the waves do not "decouple". A satisfying picture is only obtained for sufficiently small data (an assumption which ensures in particular that blowup occurs only after a partial, first order, decoupling of the waves). The proof uses crucially L^1 estimates in the space variable, which do not seem to be available in the higher dimensional case. Here again, it is possible to show that blowup occurs according to the geometric mechanism displayed in Chapter I.

It seems however reasonable to conjecture that the geometric mechanism also gives the correct description for general genuinely nonlinear systems and arbitrary "generic" data (not necessarily small).

We incorporate also in this chapter the case of rotationally invariant wave equations and data, which can be handled using essentially the same tools as in the one-dimensional case (method of characteristics, L^1 estimates).

7. In the last chapter, we first develop the approximation technique known as "nonlinear geometrical optics" in one- and multidimensional situations, restricting our attention to the Cauchy problem with small data. We explain how the "slow time " appears (sometimes many of them) in asymptotic expansions, and how the main terms, after an initial "free" period where the nonlinear interactions are not strongly felt, are governed by simple nonlinear equations (the "reduced equations"). A rather complete picture can thus be obtained for general systems in one space dimension.

In higher dimensions, we discuss only wave equations, for which relevant information can be obtained close to the boundary of the support of the solution. This technique, in addition to providing a large time description of the behavior of solutions, is a powerful tool for establishing precise lower bounds for the lifespans. We formulate these consequences and sketch their proofs.

Finally, combining nonlinear geometrical optics, geometric blowup and energy estimates, one can prove, for quasilinear wave equations, the existence of an "asymptotic lifespan" at which the second order derivatives of the solution become very large. Again, these results are beyond the scope of this volume, but we give an outline of the main steps.

CHAPTER I

The Two Basic Blowup Mechanisms

Introduction

In this chapter, we display examples of local solutions of systems or equations which blow up at a certain point.

The first class of examples (part A) involves systems of Ordinary Differential Equations (ODE) or semilinear equations or systems. In these cases, the solution u itself becomes infinite at a point x^0 by a process of "self-increase" in an influence domain leading to this point. In all the given examples, the constructed u is in fact infinite on a portion of hypersurface through x^0. We like to think of these various examples as occurrences of a single blowup mechanism which for convenience we call the ODE mechanism. It should be emphasized here that this mechanism has in general nothing to do with ODEs; the name comes only from the fact that ODEs are the simplest example.

In contrast with the first class of examples, the second class (part B) involves only quasilinear systems or equations. The solution u under consideration has a limit u^0 at the given point x^0, and only ∇u becomes infinite at this point. Typical examples of this situation are obtained when solving scalar conservation laws by the method of characteristics; blowup is then due to focusing of characteristics at x^0. Similar examples can be obtained for any quasilinear system whose principal symbol has a branch of real zeroes. The constructed u are either singular on a portion of hypersurface through x^0 or only at x^0 itself. We call the blowup mechanism at hand the "geometric blowup" mechanism, because it is independent of the existence of any "characteristics", existence which would be implicitly assumed if we used the term "focusing".

Finally, there are occurrences of blowups involving simultaneously the two mechanisms. We analyze one in part C.

A. The ODE mechanism

1. Systems of ODE

Let for simplicity $F(t, Y)$ be of class C^1 on the whole of $\mathbb{R} \times \mathbb{R}^N$, and consider the system

$$(1.1) \qquad Y' = F(t, Y), a < t < b, Y(t) \in \mathbb{R}^N.$$

The following theorem holds.

Theorem 1.1. *Let Y be a C^1 solution of (1.1) for $a < t < b$. Assume that $|Y(t)| \leq M$ for t close to b, $t < b$. Then Y can be continued as a C^1 solution of (1.1) on an interval $a < t < b + \varepsilon$ for some $\varepsilon > 0$.*

PROOF OF THEOREM 1.1. If Y is bounded, so is $F(t, Y)$, hence Y'. Thus $Y(t)$ has a limit $Y(b)$ when $t \to b$; by the equation, Y' has also a limit, and Y is of class C^1 for $a < t \leq b$. By solving (1.1) locally near b with initial data $Y(b)$, we obtain the desired continuation. $\diamond$

In other words, either "nothing happens" at b, or $|Y|$ is unbounded. A stronger description of the blowup behavior would be, according to the equation,

$$(1.2) \qquad \int^b |F(s, Y(s))| ds = \infty.$$

Of course, if $F \in C^\infty$ and Y is C^∞ for $a < t < b$ and does not blow up at b, Y can be continued beyond b as a C^∞ solution of (1.1).

Example. Consider, for $t < T$, the solution $y(t) = [(k-1)(T-t)]^{-\frac{1}{k-1}}$ ($k \geq 2$) of the equation $y' = y^k$. Note that the speed of blow up **decreases** with k.

For more complicated equations or systems, it is often possible to analyze the structure of the singularity by guessing the main singular term (say, for instance, $C(T-t)^{-\lambda}$), and then setting $y(t) = C(t)(T-t)^{-\lambda}$; this procedure leads to an equation in C with a singular point at T, which can be studied by standard techniques (see for example Wasow [Wa], Chapter 9).

2. Strictly hyperbolic semilinear systems in the plane

Let us consider in the plane with coordinates (x, t), a system of the form

(2.1) $\qquad Lu = \partial_t u + A(x, t)\partial_x u = F(x, t, u), \quad u \in \mathbb{R}^N,$

where the $(N \times N)$ matrix A and F are C^∞ functions of their arguments and A has real distinct eigenvalues $\lambda_1(x, t) < \ldots < \lambda_N(x, t)$. We assume A, F and u real.

By a smooth change of unknowns $u = P(x, t)v$, we can replace A by the diagonal matrix $P^{-1}AP$; thus, we assume A diagonal.

We define the real field L_j by

$$L_j = \partial_t + \lambda_j(x, t)\partial_x,$$

and call an integral curve of L_j a j-characteristic.

We will call a domain $D = D_{t_0}^{t_1} = \{(x, t), t_0 \le t \le t_1, \gamma_0(t) \le x \le \gamma_1(t)\}$ an **influence domain** for L of its basis $\{a \le x \le b, t = t_0\}$ if $x = \gamma_0(t)$ (resp. $x = \gamma_1(t)$) is the integral curve of L_N (resp. L_1) through the point (a, t_0) (resp. (b, t_0)). The name is justified by the fact that any j-characteristic through a point $(x, t) \in D$ meets the line $\{t = t_0\}$ at a point $(x_j(x, t), t_0)$ **inside** the basis segment.

The following Theorem is analogous to Theorem 1.1.

Theorem 2.1. *Let u be a C^∞ solution of (2.1) in $D \cap \{t < t_1\}$. Assume $|u| \le M$. Then u can be extended as a function in $C^\infty(D)$.*

Theorem 2.1 is a consequence of the following precise local existence theorem which we state and prove for convenience.

Theorem 2.2. *Let $D = D_0^T$ be an influence domain as above with basis $[a, b]$ in $\{t = 0\}$ and $u_0 \in C^\infty([a, b]), |u_0| \le M$. Set*

$$K = \{(x, t, v), t = 0, a \le x \le b, v \in [-M, +M]^N\}.$$

Choose K_0 a compact neighborhood of K and let F_0 be the supremum of $|F|$ over K_0. Then there exists $\eta > 0$, depending only on F_0, K_0 and the distance from K to the complement of K_0, such that the system (2.1) has a C^∞ solution u with initial value $u(x, 0) = u_0(x)$ in D_0^η.

PROOF OF THEOREM 2.2.

   a. Define $u^0(x, t) = u^0(x)$ as a smooth extension of u_0, and successively u^{n+1} as the (C^∞) solution in D of $Lu^{n+1} = F(x, t, u^n)$ with value u_0 on $\{t = 0\}$. Writing the i-component

$$u_i^{n+1}(x, t) = (u_0)_i(x_i(x, t)) + \int_0^t F_i(m_i(s), u^n(m_i(s)))\, ds$$

where $m_i(s)$ describes the i-characteristic through (x, t), we see that if η is such that $\eta(1 + \sqrt{N} F_0)$ is less than the distance of K to the complement of K_0, the graph of the function u^1 on D_0^η is contained in K_0, and so are the graphs of all the u^n. In particular, $|u^n(x, t)| \le M_0$.

   b. Let us pause here to emphasize that this rough way of choosing η to obtain a uniform bound on the $|u^n|$ gives relevant estimates in practical cases; for example, for the ODE $y' = y^2$ with initial value $a > 0$, we can take $K_0 = [-1, +1] \times [a - A, a + A]$, $F_0 = (a + A)^2$, hence $\eta = \frac{A}{(a+A)^2}$; maximizing on A gives $\eta = \frac{1}{4a}$, while the true value is $\frac{1}{a}$.

   c. We fix now an η as in (a) and prove that $\partial_{x,t}^\alpha u^n$ is bounded in D_0^η independently of n.

   We have first, with ∂ any x- or t-derivative,

$$L_j \partial u_j^{n+1} = -\partial \lambda_j \partial_x u_j^{n+1} + \partial F_j + F_j' \partial u^n .$$

Hence, with $M_1^n(t) = \max |\partial u^n|$ in D_0^t, we obtain

(2.3) $$M_1^{n+1}(t) \le C + C \int_0^t (M_1^{n+1}(s) + M_1^n(s))\, ds$$

with C independent of n.

   We take $M_1 \ge 2C$ such that $M_1^0(t) = M_1^0(0) \le M_1$. Assume now

$$M_1^n(t) \le M_1 \exp M_1 t$$

for some $n \ge 0$. Then (2.3) implies

(2.4) $$M_1^{n+1}(t) \le C \exp M_1 t + C \int_0^t M_1^{n+1}(s)\, ds.$$

Setting $z(t) = \exp -Ct \int_0^t M_1^{n+1}(s) ds$, we deduce from (2.4)

$$z'(t) \le C \exp(M_1 - C)t, \quad z(0) = 0,$$

hence

$$M_1^{n+1}(t) \leq \frac{CM_1}{M_1 - C} \exp M_1 t \leq M_1 \exp M_1 t$$

because $M_1 \geq 2C$. Thus $|\partial u^n| \leq M_1 \exp M_1 \eta$.

Assume now, with obvious notations, $M_k^n(t) \leq M_k \exp M_k t$ for some $k \geq 1$. We observe first that $|\partial^\alpha u^n(x,0)| \leq C_{k+1}$ for all $|\alpha| = k+1$ and all n. This is obvious if the derivatives are purely tangential; otherwise, it can be computed, using the equation, in terms of derivatives of u^n involving one less t-derivative, or of derivatives of u^{n-1} of order at most k. We can now repeat the preceding argument to obtain a uniform estimate $M_{k+1}^n(t) \leq M_{k+1} \exp M_{k+1} t$.

d. Finally, set $N^n(t) = \max |u^{n+1}(.,t) - u^n(.,t)|$; we have $N^n(t) \leq C \int_0^t N^{n-1}(s)ds$, hence $N^n(t) \leq 2M_0 C^n \frac{t^n}{n!}$ and the sequence u^n converges uniformly in D_0^η.

The limit u is the desired solution. $\diamond$

It is important to notice here that the constant M does not allow a control of the smallness of $u - u_0$; this would require comparing the values of u_0 at various nearby points, which can be as far apart as $2M$.

PROOF OF THEOREM 2.1. We apply Theorem 2.2 in D with an initial value on $t = t_2 < t_1$ equal to $u(.,t_2)$; we can take a compact K_0 independent of t_2, thus the corresponding η does not depend on t_2 either and we can choose $t_2 = t_1 - \frac{\eta}{2}$. $\diamond$

When the two curves γ_0 and γ_1 meet at one point (x_1, t_1) of D, the behavior of u in D near this point depends only on the values of u on the base of the "backward cone" D. The situation is very analogous to the ODE situation described in (1), intervals ending at b being replaced by cones of influence with vertex (x_1, t_1).

Let us remark that, in the framework of Theorem 2.1, if u is not bounded in D, then in general all components of u blow up simultaneously at points where u blows up. The reason for this can be seen in the following simple example.

Example. Consider the system

$$(\partial_t + \partial_x)u_1 = u_1^2, (\partial_t - \partial_x)u_2 = u_1$$

with $t_0 = 0, t_1 = 1, a = -1, b = +1, u_2(x,0) = 0, u_1(x,0)$ having a maximum at a with value 1. We have immediately

$$u_1(x,t) = \frac{u_1(x-t,0)}{1 - tu_1(x-t,0)};$$

thus $u_1(1-t,t)$ blows up like $\frac{C}{(1-t)}$ when $t \to 1$. Hence u_2 also blows up.

In this example, we can see that u_1 blows up "by itself", while u_2, satisfying a linear differential equation with u_1 as a source term, is forced to blow up by u_1.

3. Semilinear wave equations

The analogy observed between the one- and the two-dimensional situations handled in the previous sections is not really a surprise because they both involve integrating differential equations along curves. With the following theorem, we turn now to truly multidimensional situations.

Theorem 3.1. *Consider the wave equation*

$$(3.1) \qquad (\partial_t^2 - \Delta_x)u = \varepsilon \frac{2(m+1)}{(m-1)^2} u^m + \sum_{-\infty}^{m-1} a_j(x,t)u^j, \quad \varepsilon = \pm 1$$

where m is an integer, $m \geq 2$, the a_j' s are given analytic functions near some point $m_0 = (x_0, t_0)$ and the sum is finite. Fix S an analytic space-like hypersurface through m_0 with equation $\{t = \psi(x)\}$ (that is, $|\nabla\psi|^2 < 1$). Then, for $\varepsilon = +1$, equation (3.1) has singular solutions defined for $t > \psi(x)$ which, near m_0, blow up precisely on S, and satisfy

$$(3.2) \qquad \lim_{t \to \psi(x)} (t - \psi(x))^{\frac{2}{(m-1)}} u(x,t) = (1 - |\nabla\psi(x)|^2)^{\frac{1}{(m-1)}}.$$

Remark. *We will see that for $\varepsilon = -1$, the construction of singular solutions breaks down. In fact, for m not too large, one can prove that no blowup occurs.*

PROOF OF THEOREM 3.1.

a. Set $\tau = (t - \psi)^{\frac{1}{(m-1)}}$ and look for u of the form $u = \tau^{-2}v(x,\tau)$. By substitution in (3.1), we find an equation for v of the form

$$(3.3) (1 - |\nabla\psi|^2)(\tau\partial_\tau - 2)(\tau\partial_\tau - m - 1)v - 2\varepsilon(m+1)v^m =$$
$$= \tau g(x, \tau, v, \tau v_\tau, \nabla_x \tau v_\tau) + (m-1)^2 \tau^{2m-2} \Delta v.$$

b. Let us pause here to explain some elementary facts about **Fuchsian PDE**.

An ordinary differential equation with smooth coefficients a_k of the form

$$t^m y^{(m)} + \sum_{0 \le k \le m-1} a_k(t) t^{\mu_k} y^{(k)} = 0$$

is said to have a regular singular point at the origin $t = 0$ (or to be Fuchsian) if the weights $k - \mu_k$ of the terms $t^{\mu_k}(\frac{d}{dt})^k$ in the equation are all less than or equal to zero. Such an equation is reduced to a first order system by adapting the usual procedure and setting

$$Y_0 = y, \ Y_1 = ty', \dots, Y_{m-1} = \left(t\frac{d}{dt}\right)^{m-1} y.$$

We obtain a system

$$t\frac{d}{dt}Y + A(t)Y = 0.$$

For more details, consult [Wa], Chapter II.

Consider now an operator

$$t^m \partial_t^m + \sum_{k < m, k+|\alpha| \le m} a_{k,\alpha}(x, t)\partial_t^k \partial_x^\alpha.$$

We call it Fuchsian if we can write $a_{k,\alpha}(x, t) = t^{\mu_{k,\alpha}} b_{k,\alpha}$ with $b_{k,\alpha}$ smooth and

$$\mu_{k,\alpha} \ge k,$$

with **strict** inequality if $|\alpha| \ge 1$.

This definition is motivated by the case of ODE and by the following example: the integral curves of the field $t\partial_t + t^\mu \partial_x$ do not meet $\{t = 0\}$ (except for the x-axis itself), unless $\mu > 0$. Thus, for $\mu = 0$, the Cauchy problem is certainly not locally well-posed with respect to the initial surface $\{t = 0\}$.

Again, such an operator can be reduced to a system

$$t\partial_t Y + AY = tF(x, t, Y, \nabla_x Y).$$

For such systems, the Cauchy problem is known to be well-posed in the analytic case (see [BG]).

c. We come back now to (3.3), and see that this equation is Fuchsian.

If we try to insert $v = v_0 + \tau v_1 + \ldots$ into (3.3), we obtain first

$$(1 - |\nabla\psi|^2)v_0 - \varepsilon v_0^m = 0,$$

which gives, if $\varepsilon = +1$, $v_0 = (1 - |\nabla\psi|^2)^{\frac{1}{m-1}}$ as announced in (3.2). If $\varepsilon = -1$, only $v_0 = 0$ is a solution, and the construction does not work.

We can go on like this and compute recursively $v_1, \ldots, v_{2m+1}$, but not v_{2m+2}; in fact, if we set

$$v = \sum_0^{2m+1} v_j(x)\tau^j + \tau^{2m+2}w(x,\tau)$$

we obtain for w the equation

$$(1 - |\nabla\psi|^2)(\tau\partial_\tau)(\tau\partial_\tau + 3m + 1)w = \phi(x) + h(x,\tau,\tau w, \tau^2 w_\tau, \nabla_x\tau^2 w_\tau)$$
$$+ (m-1)^2\tau^{2m-2}\Delta w$$

where h vanishes for $\tau = 0$. This is a hint of the presence of some logarithmic terms, classical in this context.

d. If we set

$$w = \frac{\phi \ell n\tau}{(3m+1)(1 - |\nabla\psi|^2)} + z,$$

where $z = z(x, \tau, \tau\ell n\tau)$, we get an equation for z which finally can be reduced, by introducing the $n + 2$ unknowns $Y_0 = z, Y_1 = \tau\partial_\tau z, Y_{2+j} = \tau\partial_j z$, to a system

(3.4) $$NY + AY = f(x, \tau, \eta, Y, \nabla_x Y).$$

Here, $\eta = \tau\ell n\tau$, $N = \tau\partial_\tau + (\tau + \eta)\partial_\eta$, A is a constant matrix and f is analytic and vanishing for $\tau = \eta = 0$. The appearance of N comes from the fact that $\tau\partial_\tau[f(\tau, \tau\ell n\tau)] = Nf(\tau, \eta)$. In [KL], such systems are called "generalized Fuchsian" (because of the extra variable η) equations and the following theorem is proved, generalizing the theorem of [BG].

Theorem. *If A has no eigenvalue with negative real part, the system (3.4) has, near the origin, exactly one analytic solution which vanishes for $\tau = \eta = 0$.*

Applying this theorem to (3.4) yields z (depending on one arbitrary function of x), and finishes the proof.   ◇

In the previous theorem, it would be interesting to understand how much hyperbolicity from the wave equation finally remains in the system in Y, in order to solve it for smooth data.

Obviously, this kind of technique can be used to construct singular solutions in many similar situations.

B. The geometric blowup mechanism

1. Burgers' equation and the method of characteristics

a. We consider, in the plane, Burgers' equation

$$(1.1) \qquad \qquad \partial_t u + u \partial_x u = 0$$

with a smooth enough initial value $u(x,0) = u_0(x)$ defined near X^0, and $u_0'(X^0) < 0$.

A given smooth solution u of (1.1) is constant along the integral curves of $\partial_t + u \partial_x$, hence these curves are the straight lines $x = X + s u_0(X), t = s$, indexed by X and parametrized by s (this is the so-called "method of characteristics").

Taking an x-derivative of (1.1), we obtain

$$(\partial_t + u \partial_x) \partial_x u + (\partial_x u)^2 = 0.$$

Thus $\partial_x u$, restricted to the characteristic through $(X,0)$, is a function $q(s)$ which satisfies the equation $q' + q^2 = 0$, hence becomes infinite for $s^{-1} = -u_0'(X)$.

We define γ to be the locus of all these blowup points

$$\gamma = \left\{ (x,t), x = X + s u_0(X), t = s, s^{-1} = -u_0'(X) \right\}.$$

(i) Assume first $u_0''(X^0) \neq 0$: the set γ is then a smooth curve to which the characteristics are tangent. The solution u, defined near the point $m_0 = (X^0 + s_0 u_0(X^0), s_0)$ $(s_0^{-1} = -u_0'(X^0))$ on one side of γ by the fact of being constant on the characteristics, is singular everywhere on γ.

(ii) Assume now $u_0''(X^0) = 0$, $u_0'''(X^0) > 0$ (resp. < 0); the curve γ has a cusp at the point m_0, pointing downwards (resp. upwards). In the first case, the solution u is defined at least for $t < s_0$, and is singular at the cusp point; in the second case, u is defined below γ, and is singular everywhere on γ.

To prove these elementary facts, we note that by the definition of γ,

$$\left(\frac{dx}{dX}, \frac{dt}{dX}\right)(X) = \frac{u_0''}{(u_0')^2}(u_0(X), 1),$$

which proves (i). In case (ii),

$$\left(\frac{d^2x}{dX^2}, \frac{d^2t}{dX^2}\right)(X^0) = \frac{u_0'''}{(u_0')^2}(u_0(X^0), 1).$$

Remark that it is natural for γ to be the envelope of the characteristics; in fact, the solution being constant along the characteristics, blowup occurs when two infinitesimally closed characteristics meet.

b. We now **reinterpret** the facts established in (a).

Consider the functions

$$\phi(X, T) = X + T u_0(X), \quad v(X, T) = u_0(X)$$

and the mapping

(1.2) $$(X, T) \mapsto \Phi(X, T) = (x, t) = (\phi(X, T), T).$$

The characteristics are the images by Φ of the lines $X = C$ and the solution u satisfies

(1.3) $$u(\Phi(X, T)) = v(X, T).$$

The set γ is the image by Φ of the set

$$\{(X, T), \partial_X \phi(X, T) = 0\},$$

that is, the set of points where the differential Φ' is not invertible. Since

$$\partial_X v = \partial_x u(\Phi) \partial_X \phi,$$

the vanishing of $\partial_X \phi$ is precisely the reason for the blowup of $\partial_x u$ (assuming $\partial_X v \neq 0$ then). Remark finally that ϕ and v satisfy

(1.4) $$\partial_T \phi = v, \partial_T v = 0.$$

Suppose now, conversely, that we are given, near some point $m^0 = (X^0, T^0)$, a solution of (1.4), with

$$\partial_X \phi(m^0) = 0, \quad \partial_X v(m^0) \neq 0.$$

Whenever a function u can be defined satisfying (1.3), it is a solution of Burgers' equation, and $\partial_x u$ becomes infinite at $\Phi(m^0)$. For example, in the case (i) above, ϕ satisfies

$$\partial_X \phi(X^0, T^0) = 0, \quad \partial_X^2 \phi(X^0, T^0) \neq 0,$$

which means that Φ has a fold singularity at m^0. The image of Φ is one side of γ, and u can be defined on this side by

$$(1.5) \qquad\qquad u(x, t) = v(\psi(x, t)),$$

where ψ is one of the two possible choices of a point (X, T) such that $\Phi(\psi(x, t)) = (x, t)$.

In the case (ii), ϕ satisfies

$$\partial_X \phi(m^0) = 0, \quad \partial_X^2 \phi(m^0) = 0, \quad \partial_X^3 \phi(m^0) \neq 0, \quad \partial_{XT}^2 \phi(m^0) \neq 0,$$

which means that Φ has a cusp singularity at m^0. The image of Φ is a neighborhood of $\Phi(m^0)$; the points (x, t) exterior to the cusp correspond to only one point (X, T), while the points interior to the cusp correspond to three. Singular solutions u of Burgers' equation can be defined in these regions by appropriate choices of ψ.

In both cases, the choices of ψ are visualized here by the picture of the various characteristics through (x, t) tangent to γ.

c. For more general scalar equations of the form

$$(1.6) \qquad\qquad \partial_t u + \sum_{1}^{n-1} a_j(u) \partial_j u = 0,$$

the theory is completely similar, $\partial_x u$ being replaced by $\Sigma a_j'(u) \partial_j u$. Note that this quantity is the divergence d of the field

$$a(u) = (a_1(u), \ldots, a_{n-1}(u)).$$

In fact, the derivative of $d = \Sigma a_j'(u) \partial_j u$ along a characteristic is

$$\Sigma a_j'(u)(\partial_t \partial_j u + \Sigma a_k(u) \partial_k \partial_j u) = -(\Sigma a_j'(u) \partial_j u)^2.$$

2. Blowup of a quasilinear system

It turns out that the construction of singular solutions to Burgers'
equation explained in (1 b) can be extended to any quasilinear system
whose symbol has a branch of real zeroes. In this theory, the analog of
system (1.4) will be called "blowup system", and singular solutions satisfying the analog of (1.5) will be called "blowup solutions". We develop
now the general theory.

Consider a quasilinear system

$$(2.1) \qquad Lu = \sum_1^n A_j(x,u)\partial_j u + B(x,u) = 0, \quad x \in \mathbb{R}^n, \quad u \in \mathbb{R}^N,$$

where A_j and B are real smooth matrices of size $N \times N$ and $N \times 1$ defined
near (x^0, u^0).

Set $A(x,u,\xi) = \Sigma A_j(x,u)\xi_j$ and $\sigma = \det A$ the principal symbol of
the linearized operator of L on u. We make on L the following assumption:

Assumption. *There exists $\xi^0 \neq 0$ and a real simple eigenvalue $\lambda(x,u,\xi)$
of $A(x,u,\xi)$, defined near (x^0, u^0, ξ^0), such that*

$$(2.2) \qquad \lambda(x^0, u^0, \xi^0) = 0, \quad \partial_\xi \lambda(x^0, u^0, \xi^0) \neq 0.$$

We will denote by ℓ and r left and right eigenvectors of A for the
eigenvalue λ.

a. Recall first the classical definition by Lax of a genuinely nonlinear
(shortly written GNL) eigenvalue.

Definition. *A simple eigenvalue $\lambda(x,u,\xi)$ of the symbol $A(x,u,\xi)$ of the
system L is said to be genuinely nonlinear at (x^0, u^0, ξ^0) if*

$$(2.3) \qquad (r\partial_u \lambda)(x^0, u^0, \xi^0) \neq 0.$$

For instance, in the case of the scalar equation

$$\partial_t u + \mu(u)\partial_x u = 0,$$

we have $\lambda(u,\xi,\tau) = \tau + \mu(u)\xi$, and λ GNL is equivalent to $\mu'(u) \neq 0$.

For a 2×2 system in diagonal form

$$\partial_t u_1 + \lambda_1(u)\partial_x u_1 = 0, \quad \partial_t u_2 + \lambda_2(u)\partial_x u_2 = 0,$$

λ_i GNL is equivalent to $\partial_{u_i}\lambda_i \neq 0$.

b. Let now κ be some index with $\xi_\kappa^0 \neq 0$ and P a real $(N-1) \times N$ matrix such that the matrix P^0 with first line ℓ and

$$P_{k,j}^0 = P_{k-1,j}, \quad k > 1$$

is invertible (such a pair will be called "admissible").

Assume for simplicity $x^0 = 0$, and set, for $X \in \mathbb{R}^n$ close to zero and ϕ real

(2.4) $\qquad \Phi(X) = (X_1, \ldots, X_{\kappa-1}, \phi(X), X_{\kappa+1}, \ldots, X_n)$

(2.5) $\qquad \eta(X) = -\varepsilon(\partial_1\phi, \ldots, \partial_{\kappa-1}\phi, -1, \partial_{\kappa+1}\phi, \ldots, \partial_n\phi)$

where ε is the sign of ξ_κ^0.

We introduce the following definition of the blowup system of L.

Definition 2.2. *We call the blowup system of L (for the admissible pair (κ, P)) the system L_b of size $N + 1$ in the unknown (ϕ, v) given by*

$(2.5)_a \qquad \lambda(\Phi(X), v(X), \eta(X)) = 0,$

$(2.5)_b$
$${}^t\ell(\Phi(X), v(X), \eta(X))\left\{ \sum_{j \neq \kappa} A_j(\Phi(X), v(X))\partial_j v + B(\Phi(X), v(X)) \right\} = 0,$$

$(2.5)_c \quad P\left\{ \varepsilon A(\Phi(X), v(X), \eta(X))\partial_\kappa v + \left\{ \sum_{j \neq \kappa} A_j \partial_j v + B \right\}\partial_\kappa\phi \right\} = 0.$

Clearly, L_b is just L where we have formally performed the change

(2.6) $\qquad\qquad x = \Phi(X), u(\Phi(X)) = v(X),$

multiplied on the left by $\mathrm{diag}(1, \partial_\kappa\phi, \ldots, \partial_\kappa\phi)P^0$.

In the future, we will only consider solutions of L_b with the following properties:

(i) (ϕ, v) is of class C^2,

(ii) $\Phi(0) = x^0 = 0$, $v(0) = u^0$, $\eta(0) = \mu\xi^0$, $\mu > 0$,

(iii) $\partial_\kappa\phi(0) = 0$, $d(\partial_\kappa\phi) \neq 0$.

c. Examples.

(i) For Burgers' equation and κ corresponding to x, we find for L_b

$$\text{(2.7)} \qquad \partial_T \phi = v, \; \partial_T v = 0,$$

which has been already considered in (1 b).

(ii) For a system

$$\text{(2.8)} \qquad \partial_t u + A(u)\partial_x u = 0,$$

let the μ_j be the eigenvalues of A with left and right eigenvectors $\ell_j(u), r_j(u)$. The blowup corresponding to x and to the first eigenvalue $\lambda_1 = \tau + \mu_1 \xi$ yields the system L_b

(2.9)
$$\partial_T \phi = \mu_1(v), \;\; {}^t\ell_1(v)\partial_T v = 0, \;\; (\mu_k - \mu_1)(v) \; {}^t\ell_k \partial_X v + \partial_X \phi \, {}^t\ell_k \partial_T v = 0.$$

In this case, the whole procedure is just straightening out the 1-characteristics.

(iii) Let us consider the equation

$$\text{(2.10)} \quad \partial_{xt}^2 u + \partial_x u \partial_{x^2}^2 u + a \partial_{y^2}^2 u = 0, \;\; (x, y, t) \in \mathbb{R}^3, \;\; a = \text{constant}.$$

By setting $u_1 = \partial_x u, u_2 = \partial_y u$, one reduces (2.10) to a system L whose blowup system L_b is equivalent (after elimination of v_2) to

$$\partial_T \phi = v_1 + a(\partial_Y \phi)^2, \; \partial_{XT}^2 v_1 - 2a\partial_Y \phi \partial_{XY}^2 v_1$$
$$\text{(2.11)} \qquad\qquad + a\partial_X \phi \partial_{Y^2}^2 v_1 - a\partial_{Y^2}^2 \phi \partial_X v_1 = 0.$$

In all these examples, we observe the appearance of a double characteristic field: ∂_T for the first two, $\partial_T - 2a\partial_Y \phi \partial_Y$ for the last one. This will be confirmed in general by Proposition 4.1.

3. Blowup solutions

We will now associate to any solution of the blowup system L_b of a given system L one or several singular solutions u of L, according to the following definition.

Theorem and Definition 3.1. *Let (κ, P) an admissible pair for (L, λ) and (ϕ, v) a solution of the blowup system L_b. Assume that there are a* *connected open set D ($x^0 \in \bar{D}$) and a continuous map ψ from $\bar{D}$ to $\mathbb{R}^n$ with*

(3.1) $\psi(x^0) = 0,\ \Phi(\psi(x)) = x,\ \det \Phi'(\psi(x)) \neq 0,\ x \in D.$

Set then $u(x) = v(\psi(x))$, $x \in D$. The function u is a C^2 solution of L in D, called the "blowup solution". The class of all the solutions obtained in this manner does not depend on the choice of the admissible pair (κ, P).

Strictly speaking, we should say "germs of solutions" in the preceding theorem; in fact, we consider only germs (at the origin) of solutions of L_b, the corresponding u being then defined in the intersection of D with some neighborhood of x^0; moreover, one may have to again shrink these neighborhoods when comparing two solutions.

PROOF OF THEOREM 3.1.
    a. We prove first the following

Proposition. *If (κ_1, P_1) and (κ_2, P_2) are admissible pairs with corresponding systems L_b^1 and L_b^2 and (ϕ_1, v_1) is a solution of L_b^1, there is a C^2 local diffeomorphism h ($Y = h(X), h(0) = 0$) and a solution (ϕ_2, v_2) of L_b^2 such that*

$$v_2(h(X)) = v_1(X),\ \Phi_2(h(X)) = \Phi_1(X).$$

In fact, define $Y = h(X) = (h_1(X), \ldots, h_n(X))$ by

$$h_{\kappa_2}(X) = X_{\kappa_1},\ h_{\kappa_1}(X) = \phi_1(X),\ h_j(X) = X_j,\ j \neq \kappa_1, \kappa_2.$$

Because $\partial_{X_{\kappa_2}} \phi_1 \neq 0$ by assumption, h is a local C^2 diffeomorphism; the map $\Phi_2(Y) = \Phi_1(h^{-1}(Y))$ is of the form $x_{\kappa_2} = \phi_2(Y)$, $x_j = Y_j$, $j \neq \kappa_2$, with

$$Y_{\kappa_1} = \phi_1(Y_1, \ldots, Y_{\kappa_1-1}, Y_{\kappa_2}, Y_{\kappa_1+1}, \ldots, Y_{\kappa_2-1}, \phi_2, Y_{\kappa_2+1}, \ldots, Y_n).$$

We thus obtain

$$\eta_2(Y) = -\varepsilon_2(\partial_{Y_1}\phi_2, \ldots, \partial_{Y_{\kappa_2-1}}\phi_2, -1, \ldots, \partial_{Y_n}\phi_2)$$
$$= |\partial_{X_{\kappa_2}}\phi_1|^{-1} \eta_1(h^{-1}(Y)).$$

Set now $v_2(Y) = v_1(h^{-1}(Y))$; the equation

$$\lambda(\Phi_1(X), v_1(X), \eta_1(X)) = 0$$

implies

$$\lambda(\Phi_2(Y), v_2(Y), \eta_2(Y)) = 0$$

because of the positive homogeneity of λ.

Let $\bar{X}$ close to 0 with $\Phi_1'(\bar{X})$ invertible; Φ_1 is then a local diffeomorphism near $\bar{X}$, and we obtain a solution u of L near $\Phi_1(\bar{X})$ by $u(\Phi_1(X)) = v_1(X)$. Thus $v_2(Y) = u(\Phi_2(Y))$ is a solution near $\bar{Y} = h(\bar{X})$ of L_b^2; points as $\bar{Y}$ being dense near 0, (ϕ_2, v_2) is everywhere a solution of L_b^2.

b. Let now u be a blowup solution corresponding to D, ψ and a solution (ϕ_1, v_1) of L_b^1; according to (a), there exists (ϕ_2, v_2) such that, with $\bar{\psi}(x) = h(\psi(x))$, we have $\bar{\psi}(0) = 0, \Phi_2(\bar{\psi}(x)) = x, \Phi_2'(\bar{\psi}(x))$ invertible for $x \in D$, and $u(x) = v_2(\bar{\psi}_2(x))$.

Examples.

(i) Burgers' equation

The various solutions displayed in (1) for the Burgers' equation are obtained from the solution $v(X, T) = u_0(X), \phi(X, T) = X + Tu_0(X)$ of the blowup system by choosing an appropriate domain D and a "branch" ψ on D. We have then $\psi(x, t) = (X, t)$ and X represents geometrically the point of intersection with the x-axis of the chosen characteristic through (x, t).

In the first case 1 (i), among the two characteristics through (x, t) tangent to γ, we choose the one for which the contact point (x_1, t_1) satisfies $t_1 \geq t$; the other choice defines of course another local solution of Burgers' equation, corresponding to different boundary conditions.

In the case of a downward pointing cusp, there is only one possible choice, which we call the "exterior cusp solution". For an upward pointing cusp, the choice of the characteristic is the same as before, that is, the contact point is above (x, t) (we call this solution the "interior cusp solution"); the other choice yields two different possibilities, corresponding to the continuation of the exterior cusp solution (viewed now upside down) as a multivalued solution in the interior of the cusp.

(ii) Simple waves

For systems like (2.8), solutions of the form

(3.2) $$u(x, t) = v(\zeta(x, t)), \zeta \in \mathbb{R}$$

are traditionally called "simple waves" in the literature (see for instance [CFr], [Ma] or [Sm]). For u to be a solution, we need

$$v' = r_j(v), \partial_t \zeta + \mu_j(v(\zeta))\partial_x \zeta = 0$$

for some j. Thus, in addition to integrating the field $r_j(u)$ in $\mathbb{R}^N$, we just have to solve a scalar nonlinear equation on ζ.

If we consider now L_b, we see that special solutions exist of the form

$$v \equiv v(X), v'(X) \text{ colinear to } r(v(X)), \partial_T \phi = \mu_1(v(X)).$$

These solutions correspond exactly to the simple waves. We see in this way that the class of blowup solutions is a generalization of the class of simple waves. Moreover, blowup solutions exist also for higher space dimensions.

We close this section with two definitions.

Definition. *Let u be a blowup solution corresponding to (ϕ, v). If the map Φ has a fold point at the origin (that is, $\partial_\kappa^2 \phi(0) \neq 0$), we say that u is a **fold solution**; if the map Φ has a cusp point at the origin (that is, $\partial_\kappa^2 \phi(0) = 0$, $\partial_\kappa^3 \phi(0) \neq 0$ and, as we have already assumed, $d(\partial_\kappa \phi) \neq 0$), we say that u is a **cusp solution**.*

In the case of a fold solution, the image of Φ is easily seen to be one half-space limited by the characteristic fold surface S, image by Φ of the surface $\{x, \partial_\kappa \phi(x) = 0\}$; we then take automatically this half-space for D.

We could of course go on like this for arbitrary type of singularities of Φ; the classification of blowup solutions follows the classification of singularities (see for instance [GG]). We can also define a stable u as corresponding to a germ of Φ stable at the origin, etc.

We turn now to solving the blowup system of a given system.

4. How to solve the blowup system

Let us say right away that, except in the case $n = 2$ (that is, the case of one space dimension for hyperbolic systems), we do not know of a general procedure to obtain enough C^∞ solutions of the blowup system L_b of a given system L. This is due to some degeneracy of L_b at the origin, which we see by computing its principal symbol at the origin.

Proposition 4.1. *At the origin, the (scalar) principal symbol σ_b of the linearized operator of L_b on a solution (ϕ, v) is*

$$(4.1) \qquad \sigma_b(0, \zeta) = C \zeta_\kappa^{N-1} \Big(\sum_{j \neq \kappa} \partial_{\xi_j} \lambda \zeta_j \Big)^2,$$

where $C \neq 0$ and the coefficients $\partial_{\xi_j} \lambda$ are not all zero.

PROOF.

a. The last statement follows from the identity $\Sigma \xi_j \partial_{\xi_j} \lambda = 0$, $\partial_\xi \lambda \neq 0$ and $\xi_\kappa^0 \neq 0$.

b. By inspection of (2.5), we have

$$\sigma_b = (\Sigma \partial_{\xi_j} \lambda \zeta_j) \det \begin{pmatrix} {}^t\ell \Sigma A_j \zeta_j \\ \varepsilon P A \zeta_\kappa \end{pmatrix}.$$

Since A has rank $N - 1$, the lines $\ell_1, \ldots, \ell_{N-1}$ of PA are independent, and orthogonal to r. Hence we can write ${}^t\ell A_j = ({}^t\ell A_j r) r |r|^{-2} + \Sigma \alpha_j \ell_j$ and

$$\det \begin{pmatrix} {}^t\ell A_j \\ PA \end{pmatrix} = |r|^{-2} \det \begin{pmatrix} r \\ PA \end{pmatrix} ({}^t\ell A_j r).$$

On the other hand, by taking a ξ_j-derivative of the identity $A(\xi)r(\xi) = \lambda(\xi)r(\xi)$, we obtain ${}^t\ell A_j r = \partial_{\xi_j}\lambda\, {}^t\ell r$; finally, ${}^t\ell r \neq 0$ because λ is simple. $\Diamond$

In the case of analytic coefficients, we can use the Cauchy-Kovalevsky theorem with respect to any noncharacteristic hypersurface to obtain solutions of L_b, as explained in the following theorem.

Theorem 4.2. *Assume that the coefficients A_j and B of L are analytic near (x^0, u^0). We can choose $\partial_X v(0)$ such that (2.5) is satisfied at $X = 0$, by taking $\partial_{X_\kappa} v(0)$ colinear to r and the other $\partial_{X_j} v(0)$ appropriately chosen. For any noncharacteristic δ (that is, $\delta_\kappa \neq 0$, $\sum_{j \neq \kappa} \partial_{\xi_j} \lambda(x^0, u^0,$*

$\xi^0)\delta_j \neq 0)$, we choose new orthogonal coordinates $(Y_1, \ldots, Y_n)$ for which δ is colinear to the Y_1-axis. The system L_b is then equivalent (for functions with the chosen jets at the origin) to a system

$$(4.2) \qquad \partial_{Y_1}(\phi, v) = F(Y, \phi, v, \partial_{Y'}\phi, \partial_{Y'}v)$$

where Y' denotes the $Y_j, j \geq 2$; we can solve this system for any analytic data (ϕ_0, v_0) given on $\{Y_1 = 0\}$. Moreover,

  (i) we can choose the 2-jet of ϕ_0 at 0 to obtain a fold solution.

 (ii) If $n \geq 3$, we can choose the 3-jet of ϕ_0 at 0 to obtain a cusp solution.

(iii) If $n = 2$, we can choose the 3-jet of ϕ_0 at 0 to obtain a cusp solution if and only if λ is genuinely nonlinear.

The proof of the theorem is obvious, except for the last three technical statements, for which we refer to [Al8].

In other words, we can use the Cauchy-Kovalevsky theorem to obtain solutions of L_b for which Φ displays the various types of singularities we are interested in.

The difference between (ii) and (iii) in Theorem 4.2 is easy to understand: for a **linear** system, we can obtain a cusp solution by "propagating" a cusped (in dimension $n - 1$) initial datum; this is not possible for $n = 2$.

Let us finally remark that in some cases, only partial analyticity is required for solving L_b; this is the case in Example (iii) of (2), for instance (see [Al8]).

5. How ∇u blows up

The blowup solution u itself is, by construction, continuous on $\bar{D}$; only its gradient blows up at x^0, in a way indicated in the following theorem.

Theorem 5.1. *Let u be a blowup solution of L corresponding to a solution (ϕ, v) of L_b for which $\partial_\kappa v(0) \neq 0$. Then, for $x \in D$,*

$$(5.1) \qquad \partial_x u(x) = C(x)(\partial_\kappa \phi)^{-1}(\psi(x))r(x, u(x), \xi(x))^t\xi(x) + R(x),$$

where $C(x)$ and $R(x)$ are continuous on $\bar{D}$, $C(x^0) \neq 0$ and $\xi(x) = \eta(\psi(x))$.

PROOF OF THEOREM 5.1. Let $\kappa = 1$ for simplicity. The function v being a solution of (2.5), we have $\partial_1 v = \alpha r + \partial_1 \phi w$ for some w and $\alpha \neq 0$. Moreover, we have the relations

$$\partial_1 u = (\partial_1 \phi)^{-1} \partial_1 v, \partial_j u = \partial_j v - (\partial_1 \phi)^{-1} \partial_j \phi \partial_1 v.$$

This yields (5.1).   ◇

For instance, if u is a blowup solution of (2.8), because of (2.9), we have

$$^t\ell_k \partial_x u = -(\mu_k - \mu_1)^{-1} \, ^t\ell_k \partial_T v, \quad k \neq 1,$$

which shows that only the component $^t\ell_1 \partial_x u$ can blow up.

In particular, the formula (5.1) reduces the question "how does ∇u become infinite when x approaches x^0 ?" to a purely geometric problem on the behavior of the factor $[\partial_\kappa \phi(\psi(x))]^{-1}$ in D close to x^0.

For a fold solution, this factor is easily seen to behave like $[d(x)]^{-\frac{1}{2}}$, where $d(x)$ is the distance from x to the fold surface. For an exterior cusp solution, the rate depends on the direction of approach to x^0 (see for instance Chapter IV, 1 or [Le]).

6. Singular solutions and explosive solutions

Up to now, we have made no distinction between linear and nonlinear cases.

If L is a linear system, the blowup solutions that we have considered are singular on (smooth or not) characteristic hypersurfaces. In particular, fold solutions are singular on the smooth characteristic fold surface S, a configuration easy to obtain by the methods of linear geometrical optics; the bicharacteristic strip through (x^0, ξ^0) (let us recall that ξ^0 is normal to S at x^0) is entirely contained in S. Similarly, cusp solutions can be obtained by "propagating" cusp data, and so on.

For a nonlinear system, the situation is completely different: roughly speaking, one can say that the singularities of the blowup solution are **created**, and not "propagated".

To give a precise meaning to this statement, we have to study the behavior of the characteristic curves of the linearized operator of L on u leading to x^0. When these curves are in D, they are obtained as images (by

Φ) of characteristic curves of the part of L_b corresponding to equations $(2.5)_b$, $(2.5)_c$ (linearized in v only), the symbol of which is, according to the proof of Proposition 4.1, $\tilde{\sigma}_b = C\zeta_\kappa^{N-1} \sum_{j\neq\kappa} \partial_{\xi_j}\lambda\zeta_j$. We will not give here a complete study of these curves, referring to [Al8] for details. We give only the result for the special family of the bicharacteristic strips issued from points of the form $(0, \zeta^0), \zeta_\kappa^0 \neq 0$.

Proposition 6.1. *Let (ϕ, v) be a solution of L_b with $\partial_\kappa v(0) \neq 0$. Denote by $\mathcal{C}(\zeta^0) = (X(s), \zeta(s))$, s close to 0, the bicharacteristic strip of $\tilde{\sigma}_b$ issued from $(0, \zeta^0)$ where ζ^0 satisfies*

$$(6.1) \qquad \zeta_\kappa^0 \neq 0, \quad \sum_{j\neq\kappa} \partial_{\xi_j}\lambda(x^0, u^0, \xi^0)\zeta_j^0 = 0.$$

(i) *Assume that Φ has a fold and that λ is GNL. Then the image $\Phi(X(s))$ is a smooth curve, tangent to the fold surface for $s = 0$ with a **contact of order exactly two**. The image of each of the two half-curves $\mathcal{C}_\pm(\zeta^0) = \{(X(s), \zeta(s)), \pm s > 0\}$ is an arc of bicharacteristic $(x(s), \xi(s))$ (above D) of the linearized operator of L on a blow up solution u corresponding to (ϕ, v), the two half-curves corresponding to the two different choices of ψ. Moreover, the projections $(x(s), \frac{\xi(s)}{|\xi(s)|})$ of these arcs on the cosphere bundle tend to one of the points $(x^0, \pm\frac{\xi^0}{|\xi^0|})$.*

(ii) *Assume that Φ has a cusp and that λ is GNL. Then the image $\Phi(X(s))$ is a smoth curve crossing the cusp and **transverse** to the edge of the cusp. The image of each of the two half-curves $\pm(\zeta^0)$ is an arc of bicharacteristic (above D) of the linearized operator of L on the exterior and interior cusp solution corresponding to (ϕ, v). Moreover, the projections of these arcs on the cosphere bundle tend to one of the points $(x^0, \pm\frac{\xi^0}{|\xi^0|})$.*

(iii) *In all cases, the tangent to $\Phi(X(s))$ at $s = 0$ is the characteristic issued from (x^0, ξ^0) for the operator L frozen at (x^0, u^0).*

C. Combinations of the two mechanisms

In this section, we will only consider the very simple example of Burgers' equation with a nonlinear zero order term

(1.1) $\partial_t u + u \partial_x u = f(u),$

where $f(u)$ is a given real function.

1. Which mechanism takes place first?

Let u be a smooth solution of (1.1) defined for $t < T_0$. Along each integral curve of $L = \partial_t + u \partial_x$, (1.1) reduces to a nonlinear ODE of the type studied in A, and u may blow up at time T_0 on one of these curves. On the other hand, there is no reason why the nonlinearity $f(u)$ should prevent the focusing of characteristics studied in B; thus $\partial_x u$ may blow up at some point when $t \to T_0$.

Can both mechanisms take place simultaneously, or which one occurs first?

A first result is given by the following proposition.

Proposition 1.1. *Consider the solution u of (1.1) with $f(u) = u^2$ and initial value $u_0 \in C_0^2$. Assume that u_0 reaches its maximum at a point X^0 where $u_0 > 0$, $u_0'' \neq 0$. Then, if u exists for $t < T_0$, u remains bounded for $t < T_0$.*

PROOF OF PROPOSITION 1.1. If we define $\phi(X, T)$ for $t < T_0$ to be the abscissa of the point of ordinate T on the integral curve of L starting from $(X, 0)$ and set $v(X, T) = u(\phi(X, T), T)$, then (ϕ, v) is a solution of the blowup system

$$\partial_T \phi = v, \ \partial_T v = f(v), \ \phi(X, 0) = X, \ v(X, 0) = u_0(X).$$

Let $T_1 = (\max u_0)^{-1}$; since

$$v(X, T) = \frac{u_0(X)}{1 - T u_0(X)}$$

cannot live longer than T_1, $T_0 \leq T_1$. We have also $\phi(X, T) = X - \ell n(1 - T u_0(X))$.

Suppose now $T_0 = T_1$, and let X^0 a point for which $u_0(X^0) = T_1^{-1}$, $u_0''(X^0) \neq 0$; we claim that there are points close to (X^0, T_0) at which $\partial_X \phi = 0$, $\partial_X v \neq 0$. In fact, for $T < T_0$, $\partial_X \phi = 0$ is equivalent to $(u_0 - u_0')(X)T = 1$; by the implicit function theorem, there is a curve $T = h(X)$ through (X^0, T_0) of such points, with $h'(X^0) = T_0^2 u_0''(X^0) \neq 0$. Hence the curve contains points with $T < T_0$ and blowup of $\partial_x u$ occurs before T_0. $\diamond$

If we anticipate Chapter II a little and think of T_0 as being the maximal time of existence of u, by a general theorem that we will prove also in Chapter II (Theorem 2.3), we see that $\partial_x u$ cannot remain bounded for $T < T_0$ (otherwise, there would be no blowup at time T_0). We can then conclude roughly that, if the nonlinearity f is not too strong, focusing always occurs first.

2. Simultaneous occurrence of the two mechanisms

If the nonlinearity f is strong enough, both mechanisms can take place simultaneously, as we see in the following example.

Proposition 2.1. *Consider the solution u of (1.1) with $f(u) = u^4$ and initial value $u_0 \in C_0^2$. We can choose u_0 such that u exists for $t < T_0$ and $\max |u(., t)| \to \infty$ when $t \to T_0$.*

PROOF OF PROPOSITION 2.1. We have here

$$v(X, T) = u_0(X, T)(1 - 3Tu_0^3(X, T))^{-\frac{1}{3}},$$

$$\partial_X \phi(X, T) = 1 + u_0' u_0^{-3}[(1 - 3Tu_0^3)^{-\frac{1}{3}} - 1].$$

We take $T_0^{-1} = \max 3u_0^3$, and we show that we can choose u_0 such that $\partial_X \phi \neq 0$ for $t < T_0$. We take for simplicity $u_0(X) = w(\varepsilon X)$, where $w \geq 0$ has a unique positive quadratic maximum at 0 and supp $w = [a, b]$; for $\varepsilon X \notin [a, b]$, $\partial_X \phi = 1$; thus it is enough to prove $-u_0' \leq (1 - 3T_0 u_0^3)^{\frac{1}{3}}$ everywhere, because $b \geq 1 - (1 - b)^{\frac{1}{3}}$ for $0 \leq b \leq 1$. The inequality is true for $\varepsilon|X| \leq \alpha$ since the righthand side is then equivalent to $|\varepsilon X|^{\frac{2}{3}}$; elsewhere the inequality is true for ε small enough. $\diamond$

It does not seem that this sort of blowup solution can be obtained by the *Ansatz* techniques of A.3.

We believe that the two mechanisms considered in this chapter are involved in all "generic" blowups. According to Proposition 1.1, it would seem that the geometric mechanism occurs first for homogeneous ($B \equiv 0$) systems or small data. The semilinear cases, where this "primary" mechanism is suppressed, can then be viewed as linearly degenerate cases for which only the "secondary" ODE mechanism may cause blowup.

Notes

The bibliographical references for this chapter are few.

The statements about ODEs can be found in any textbook, for instance Coddington and Levinson [CL], while asymptotic expansions near a point are discussed in Wasow [Wa].

The discussion of semilinear wave equations is taken from Kichenassamy and Littman [KL], which seems to be the first case of a systematic approach. Related results may be found in Leichtnam [Lei].

The material for Section B comes essentially from Alinhac [Al8]. The special case of simple waves is crucially used in the theory of systems of conservation laws in one space dimension (see for instance Courant and Friedrichs [CFr], Lax [La2] or Smoller [Sm]). The concept of genuinely nonlinear eigenvalue is introduced in Lax [La2], and an approach of simple waves specifically related to it and to blowup can be found in Majda [Ma] and also in John [Jo6].

Some indications about the role of dissipative terms in quasilinear systems are given in [Na].

CHAPTER II

First Concepts on Global
Hyperbolic Cauchy Problems

Introduction

a. We consider here for simplicity quasilinear hyperbolic systems of the form

$$(1.1) \qquad Lu = \partial_t u + \sum_{j=1}^{n} A_j(u)\partial_j u + B(u) = 0, \quad B(0) = 0,$$

for $x = (x_1, \ldots, x_n) \in \mathbb{R}^n$, $x_0 = t \in [0, T]$, $u \in \mathbb{R}^N$. The coefficients A_j, B are assumed to be real and smooth functions of u in an open domain G containing the origin in its interior. All the solutions we consider will be classical (i.e. C^1) solutions.

For such systems, we always make the following **hyperbolicity assumption**: L is symmetrizable hyperbolic, that is, there exists on G a symmetric positive definite matrix $S(u)$ such that all the matrices SA_j are symmetric.

We do not consider general stricly hyperbolic systems since the natural examples are symmetrizable and the proofs are more delicate.

All we have to say will apply as well to scalar equations; in particular we will often consider quasilinear wave equations of the form

$$\partial_t^2 u + \sum_{i,j\geq 0} g_{ij}(u, \nabla u)\partial_{ij}^2 u + F(u, \nabla u) = 0,$$

$$(1.2) \qquad\qquad \nabla u(x) = (\partial_0 u, \partial_1 u, \ldots, \partial_n u), \ x_0 = t.$$

These equations will always be assumed strictly hyperbolic.

By a global Cauchy problem we mean that the initial value of u on $\{t = 0\}$ is a given function $u_0 \in C_0^\infty$. We make this choice for simplicity, though for most discussions, enough smoothness and decay at infinity would suffice. We dropped dependence of the coefficients on (x, t) to obtain simpler statements in these global situations.

We do not consider here the case of periodic Cauchy data, for which we refer to [GL] and [LFX].

b. This chapter is devoted to the very first concept which can be introduced for global Cauchy problems: the lifespan. This is the time $\bar{T}$ up to which a given solution retains its initial regularity. The key fact here is that an initially smooth enough solution does not undergo a gradual loss of its regularity; it keeps its initial smoothness up to $\bar{T}$, and undergoes a brutal change at time $\bar{T}$, where either the solution or its gradient (or both) become infinite. This fact can be viewed as a nonlinear propagation of regularity for the Cauchy problem. Traditionally, the brutal change is described in the literature under the name "blowup criterion".

A large number of ingenious methods have been used to test whether or not $\bar{T} < \infty$; they all make use of appropriate sign and structure (convexity for example) of the nonlinearity. We distinguish here for clarity between "functional methods" and "averaging and comparison methods". These methods, besides proving $\bar{T} < \infty$, provide upper bounds for $\bar{T}$. However, they do not display in general the correct blowup mechanisms; thus the obtained upper bounds are very rough and much larger than the true ones.

We will study in the next chapters cases where more precise information can be obtained.

1. Short time existence

We will denote by $H^s = H^s(\mathbb{R}^n)$ the usual Sobolev space with norm $|.|_s$ and by C^k the space of k-times continuously differentiable functions; the L^∞ norm will be denoted by $||.||_\infty$. For vectors $u \in (H^s)^N$, we will write simply $u \in H^s$, etc.

The following theorem is of constant use.

Theorem 1.1. *Consider a quasilinear system (1.1), which is assumed to be symmetrizable hyperbolic. For some integer $s > \frac{n}{2} + 1$, assume that*

$$(1.3) \qquad\qquad u_0 \in H^s, \quad u_0(x) \in G_0,$$

where G_0 a relatively compact subset of G. Then, for any compact neighborhood G_1 of G_0 contained in G and any $M \geq |u_0|_s$, there exists $T > 0$, *depending only on M and on G_1, and a unique u solution of (1.1) for $0 \leq t \leq T$ satisfying*

(1.4)
$$u \in C^0([0,T], H^s) \cap C^1([0,T], H^{s-1}), \quad u(x,t) \in G_1, \quad u(x,0) = u_0(x).$$

For symmetric hyperbolic systems, we refer to [Ma] for a proof; for strictly hyperbolic systems or equations, the corresponding statement is also true, and we refer to [Me]. Note in particular that solutions satisfying (1.4) are of class C^1 ("classical solutions").

2. Lifespan and Blowup Criterion

As a consequence of Theorem 1.1, we can define the lifespan of a solution.

Definition 2.1. *Consider a solution u of (1.1) with initial value $u_0 \in H^s$, as in Theorem 1.1. The lifespan T_s of u will be defined as the supremum of all $T > 0$ such that u exists for $0 \leq t \leq T$ and*

(2.1)
$$u \in C^0([0,T], H^s) \cap C^1([0,T], H^{s-1}).$$

Note that $T_s > 0$ and, possibly, $T_s = +\infty$. For $\frac{n}{2} + 1 < s' \leq s$, we have by definition $T_{s'} \geq T_s$; in other words, u could live longer if we are less demanding on its regularity. One could thus imagine a given solution becoming worse and worse with time. The essential fact is that **this cannot happen.**

Theorem and Definition 2.2. *Consider $u_0 \in C_0^\infty$. For any integer $s > \frac{n}{2} + 1$, the unique solution u of (1.1) with initial value u_0 lives for $0 \leq t < T_s$ with the regularity (2.1). In fact, T_s is a constant in s which we denote by $\bar{T}$ and call simply the **lifespan** of u.*

Note that in the situation of Theorem 2.2, the solution u is C^∞ for $t < \bar{T}$.

Example: The lifespan for Burgers' equation

Consider a solution u of Burgers' equation (Chapter I, B 1, B. 2) with initial value $u_0 \in C_0^\infty$. Then

$$\bar{T} = (\max -u_0')^{-1}.$$

Remark that the support assumption on u_0 implies $\bar{T} < \infty$ (unless of course $u_0 \equiv 0$).

Define $T_0 = (\max -u_0')^{-1}$; as explained in B. 1, $\bar{T}$ cannot be bigger than T_0. On the other hand, for the solution (ϕ, v) of the blowup system (2.7), one has $\partial_X \phi(X, T) > 0$ for $T < T_0$; for fixed $T < T_0$, the map

$$X \mapsto x = \phi(X, T)$$

is an increasing bijection from $\mathbb{R}$ onto itself, hence Φ is a C^∞ diffeomorphism of $T < T_0$ onto itself and the corresponding u is C^∞ for $T < T_0$.

$$\diamondsuit$$

Theorem 2.2 is a consequence of the more precise

Theorem 2.3. *Consider u a C^1 solution of (1.1) for $0 \leq t < T < \infty$ with initial value $u_0 \in H^s$ (s integer, $s > \frac{n}{2} + 1$). Assume that there exist a constant M and a relatively compact subset G_1 of G such that, for $0 \leq t < T$,*

$$|\nabla u(x, t)| \leq M, \quad u(x, t) \in G_1.$$

Then

$$(2.2) \qquad u \in C([0, T'], H^s) \cap C^1([0, T'], H^{s-1}), \quad T' < T$$

and, with C depending only on s, M and G_1,

$$(2.3) \qquad |u(., t)|_s \leq C|u_0|_s, \quad t < T.$$

PROOF OF THEOREM 2.3.

a. Let T_s be the lifespan corresponding to u_0. For $t < T_s$ and $|\alpha| \leq s$, we consider $u_\alpha = \partial_x^\alpha u$, which is a solution of $(\partial_t + \Sigma A_j(u)\partial_j)u_\alpha = F_\alpha$ with

$$(2.4) \qquad F_\alpha = -\sum_{j, \beta \leq \alpha, \beta \neq 0} C_\alpha^\beta \partial_x^\beta (A_j(u)) \partial_x^{\alpha-\beta} \partial_j u - \partial^\alpha (B(u)).$$

b. To evaluate the L^2 norm of F_α, we use the two following propositions:

Proposition 1. *If $f : \mathbb{R}^N \to \mathbb{R}$ is a smooth function with $f(0) = 0$, then, for $s > 0$, $f(u) \in L^\infty \cap H^s$ if $u \in L^\infty \cap H^s$ and*

(2.5) $$|f(u)|_s \le C|u|_s$$

for a constant C depending only on f, s and $||u||_\infty$.

Proposition 2. *If $u, v \in L^\infty \cap H^s$ for a positive integer s, then for all $\alpha, \beta, |\alpha| + |\beta| = s$,*

(2.6) $$|(\partial^\alpha u)(\partial^\beta v)|_0 \le C_s(||u||_\infty |v|_s + ||v||_\infty |u|_s).$$

These two propositions can be proved in at least two ways:

(i) One can use the Gagliardo-Nirenberg inequality (as in [Ma], Proposition 2.1); for $w \in L^\infty \cap H^s$, s being a positive integer,

(2.7) $$||\partial^\alpha w||_{L^{\frac{2s}{k}}} \le C_s ||w||_\infty^{1-(\frac{k}{s})} |w|_s^{\frac{k}{s}}, \quad k = |\alpha|, \quad 0 \le k \le s.$$

(ii) One can also use elementary properties of the Littlewood-Paley decomposition, as in [AG], Chapter II.

c. Using Proposition 2 for $s - 1$, we obtain

$$|\partial^\beta (A_j(u) - A_j(0))\partial^{\alpha-\beta}\partial_j u|_0 \le C|u|_{\text{Lip}}(|u|_s + |A_j(u) - A_j(0)|_s),$$

where $|u|_{\text{Lip}} = ||u||_\infty + ||\nabla u||_\infty$ denotes the Lipschitz norm; Proposition 1 gives then $|A_j(u) - A_j(0)|_s \le C|u|_s$, so finally

(2.8) $$|F_\alpha|_0 \le C|u|_s.$$

d. An important fact is that for symmetric systems, the constants in the L^2 energy inequality only involve the Lipschitz norm of u. Applying this inequality to u_α and summing over α, we obtain

(2.9) $$|u(.,t)|_s \le C \Sigma |u_\alpha(.,t)|_0 \le C\left(|u_0|_s + \int_0^t |u(.,\tau)|_s d\tau\right).$$

e. Let us recall the classical Gronwall Lemma.

Gronwall Lemma. *Let $\phi \in C^0([0, a))$ satisfy*

$$\phi(t) \le A + B \int_0^t \phi(s)ds.$$

Then $\phi(t) \le A \exp Bt$.

Applying this lemma to (2.9) gives (2.3) for $t < T_s$.

f. Finally, assume $T_s < T$; using Theorem 1.1 with initial value $u(.,\theta)$ on $t = \theta$ for $\theta < T_s$ close enough to T_s yields a contradiction with the definition of T_s.   ◇

PROOF OF THEOREM 2.2. Let $s > s' > \frac{n}{2} + 1$; choose $T', T' < T_{s'}$; by definition of $T_{s'}$, the hypotheses of Theorem 2.3 are satisfied with $T = T'$. Thus $T_s \geq T'$, and $T_s = T_{s'}$.   ◇

Let us remark that in the situation of Theorem 2.3, the solution u can in fact be continued after T. Thus we can rephrase Theorem 2.3 in the following way.

Blowup criterion. *Let $\bar{T} < \infty$ be the lifespan of a solution u of (1.1) with initial value $u_0 \in C_0^\infty$. Then, either*

(i) $\lim \sup \|\nabla u(.,t)\|_\infty = +\infty$ as $t \to \bar{T}$ *or*

(ii) $u(x,t)$ *is not contained in any compact subset of G for $t < \bar{T}$.*

The conclusion of this section is that there is no gradual worsening of a solution u, but a **brutal change** in regularity at time $\bar{T}$, of one of the two types (i) or (ii) described above. We have seen in the examples of Chapter I that actually one of the two or both may occur.

Remark finally that Theorems 2.2 and 2.3 do not seem to have analogue in the C^k category. Hence the way the regularity of the solution, described in this category, changes with time, is not clear (see however Theorem 3.1 of Chapter III).

3. Blowup or not? Functional methods

Consider a quasilinear system (1.1) and a compactly supported smooth initial value u_0. The first question to answer seems to be: is the lifespan $\bar{T}$ of the corresponding solution u finite or not?

Many methods have been proposed to answer this question in a rough way; we will call them rather vaguely "functional and comparison" methods. Let us start with a very simple example.

a. A functional method for Burgers' equation

Let u be the solution of Burgers' equation with $u_0 \in C_0^\infty([a, b])$. For $t < \bar{T}$, $u(.,t) \in C_0^\infty([a, b])$, because the characteristics starting outside

$[a, b]$ are vertical lines on which $u = 0$. Choose $\phi \in C^\infty([a, b])$, $\phi' > 0$ and set

$$F(t) = \int \phi(x)u(x, t)dx.$$

Using the equation, we obtain

$$F'(t) = \frac{1}{2} \int \phi'(x)u^2(x, t)dx.$$

On the other hand, by the Cauchy-Schwarz inequality,

$$F^2(t) \leq C \int \phi'(x)u^2(x, t)dx \leq 2CF'(t), \quad C = \int_a^b \frac{\phi^2(x)}{\phi'(x)}dx.$$

Thus, if $F(0) = \int \phi(x)u_0(x)dx > 0$, F cannot remain smooth beyond $T_0 = \frac{2C}{F(0)}$, hence $\bar{T} \leq T_0$.

This example is typical of the **functional method**, which we can describe roughly as follows:

(i) An appropriate functional F of u, depending on time, is introduced.

(ii) A (first or second order) differential inequality is proved on F, implying finite time blowup for well chosen data.

This is of course a very old method, and one can consult [Le1] for a survey. The remarkable fact about the above example is this: we know that $\bar{T} = (\max -u_0')^{-1}$ and that, for $t \to \bar{T}$, u remains bounded and only $\partial_x u$ becomes infinite; thus, the predicted blowup of F in fact **never** occurs. The blowup mechanism hinted at in the proof is not the correct one, and the estimate T_0 of the lifespan is not sharp.

Unfortunately, one is forced to recognize that very often, functional methods present the above mentioned drawbacks. To evaluate the efficiency of a given method, we suggest considering the following criteria:

(i) Is it possible to obtain from the method a reasonably explicit upper bound T_0 for the lifespan? If no, we call this method a "yes or not" method.

(ii) Assuming that we are considering a case where blowup can occur even for arbitrary small data say, of size ε (a case to which we will return in subsequent chapters), we can distinguish three situations:

a) The obtained $T_0(\varepsilon)$ is much bigger than the correct $\bar{T}(\varepsilon)$ ("rough" method).

b) The obtained $T_0(\varepsilon)$ has the correct order of magnitude, that is, $T_0(\varepsilon) \leq C\bar{T}(\varepsilon)$ ("semisharp" method).

c) The obtained $T_0(\varepsilon)$ is equivalent to the true one ("sharp" method).

(iii) Does the method display the correct mechanism?

For instance, the functional method proposed in the example is semisharp (because, in one space dimension, any method gives the magnitude ε^{-1} for T_0).

We present now two classical situations where a functional method is used.

b. Semilinear wave equation

Consider, in three space dimensions, the semilinear wave equation

$$(3.1) \qquad\qquad \partial_t^2 u - \Delta_x u = |u|^p.$$

Theorem 3.1. *Let u be a solution of (3.1) with initial data*

$$u(x,0) = f(x), \ \partial_t u(x,0) = g(x), \ \ f, g \in C_0^\infty$$

supported in $|x| \leq R$. Assume $1 < p < 1 + \sqrt{2}$ and $\int g \, dx > 0$. Then $\bar{T} < \infty$.

PROOF OF THEOREM 3.1.

a. Set $F(t) = \int u(x,t)dx$; using the equation and the fact that the support of u is contained in $|x| \leq R+t$, we have $F''(t) = \int |u|^p dx$; hence, by the Hölder inequality applied to the integral defining F,

$$(3.2) \qquad\qquad F''(t) \geq (t+R)^{-3(p-1)}|F|^p.$$

We make now use of the two well known facts:

(i) In dimension three, the elementary solution of the wave operator is a positive measure.

(ii) This measure is supported by the boundary $t = |x|$ of the light cone, so that the **strong Huygens principle** is valid: the free solution u_0 of $\partial_t^2 u_0 - \Delta u_0 = 0$ with data f, g is supported in the set $\{(x,t), t - R \leq |x| \leq t + R\}$.

By (i), we have $u(x,t) \geq u_0(x,t)$; by (ii), we can write

$$F_0(t) = \int u_0(x,t)dx = \int_{t-R \leq |x| \leq t+R} u_0(x,t)dx \leq \int_{t-R \leq |x| \leq t+R} u(x,t)dx$$

$$\leq C(t+R)^{\frac{2(p-1)}{p}} \left(\int |u(x,t)|^p dx \right)^{\frac{1}{p}} \leq C(t+R)^{\frac{2(p-1)}{p}} [F'']^{\frac{1}{p}}.$$

But F_0 can be computed explicitly because

$$F_0''(t) = 0, \quad F_0(0) = C_f = \int f(x)dx, \quad F'(0) = C_g = \int g(x)dx.$$

We get $F_0(t) = C_f + tC_g$, hence, if $t \geq t_0$, because of our assumption $C_g > 0$,

$$F_0(t) \geq C(t+R), \quad C > 0.$$

Finally, we have obtained

(3.3) $$F''(t) \geq C(t+R)^{2-p}, \quad C > 0, \quad t \geq t_0,$$

which implies

$$F(t) \geq C(t+R)^{4-p}, \quad C > 0, \quad t \geq t_0.$$

b. We use now the following elementary ODE lemma.

Lemma. *Consider $F(t) \in C^2[0,b)$ and suppose*

$$F(t) \geq C_0(t+R)^r, \quad F''(t) \geq C_1(t+R)^{-q}F(t)^p, \quad t \geq 0, \quad C_0, \ C_1, \ R > 0.$$

If $p > 1$, $r \geq 1$, $(p-1)r > q-2$, then b must be finite.

Here, $r = 4 - p$, $q = 3(p-1)$, thus the conditions read $p > 1$, $p^2 - 2p - 1 < 0$, hence $1 < p < 1 + \sqrt{2}$. $\diamond$

The value $1 + \sqrt{2}$ in Theorem 3.1 is not a technical artifact: it cannot be improved, as shown in [Jo6].

c. The Euler system

The motion of a polytropic, ideal compressible gas is described by the Euler system

(3.4)$_a$ $$\partial_t \rho + div(\rho u) = 0,$$

$(3.4)_b$ $$\partial_t u + (u.\nabla)u + \rho^{-1}\nabla p = 0,$$

$(3.4)_c$ $$\partial_t S + (u.\nabla)S = 0,$$

where ρ is the density of the gas, $u \in \mathbb{R}^3$ its velocity ($u.\nabla = \Sigma u_j \partial_{x_j}$), S its entropy and p its pressure given by $p = A\rho^\gamma \exp S$ ($A > 0, \gamma > 1$).

For some physical background, see [CFr].

This system can be written as a hyperbolic symmetrizable system to which the conside-rations of (1) and (2) apply (see for instance [Ma]).

For given $\bar\rho > 0$, $\bar S$, $\bar p = A\bar\rho^\gamma \exp \bar S$, consider initial data

$$\rho(x,0) = \rho^0(x) = \bar\rho + \rho^1(x) > 0, \quad u(x,0) = u^0(x),$$
(3.4) $$\quad S(x,0) = S^0(x) = \bar S + S^1(x),$$

where ρ^1, u^0 and S^1 are smooth and supported in the ball of radius R, and denote by $\bar T$ the lifespan of the smooth solution. For $t < \bar T$, the support of the disturbance ρ^1, u^0, S^1 propagates with speed at most $\sigma = (\partial_\rho p(\bar\rho, \bar S))^{\frac{1}{2}}$ (σ is called the sound speed). A proof of this essential fact can be found for instance in [Jo3]. This means more precisely that $(\rho(.,t), u(.,t), S(.,t)) = (\bar\rho, 0, \bar S)$ outside $B(t) = \{x, |x| \le R + \sigma t\}$.

Define the functionals

$(3.5)_a$ $$m(t) = \int (\rho(x,t) - \bar\rho)dx,$$

$(3.5)_b$ $$\eta(t) = \int \left[\rho(x,t) \exp \frac{S(x,t)}{\gamma} - \bar\rho \exp \frac{\bar S}{\gamma}\right] dx,$$

$(3.5)_c$ $$F(t) = \int x\rho u(x,t)dx.$$

Theorem 3.1. *If*

(3.6) $$m(0) \ge 0, \quad \eta(0) \ge 0, \quad F(0) \ge \alpha\sigma R^4 \max \rho^0(x), \quad \alpha = \frac{16\pi}{3},$$

then $\bar T < \infty$.

PROOF OF THEOREM 3.1.

a. We have

$$m'(t) = -\int \text{div}(\rho u)dx = 0$$

and

$$\eta'(t) = -\int \operatorname{div}(\rho u \exp \frac{S}{\gamma}) dx = 0,$$

hence $m(t) = m(0), \eta(t) = \eta(0)$. Moreover,

$$F'(t) = \int x \, \partial_t(\rho u) \, dx = -\int \operatorname{div}\left[(xu)\rho u + x(p - \bar{p})\right]$$
$$+ \int \left[\rho |u|^2 + 3(p - \bar{p})\right] dx = \int \left[\rho |u|^2 + 3(p - \bar{p})\right] dx.$$

b. Now $\int (p - \bar{p}) dx \geq 0$. In fact, $(p - \bar{p})(., t)$ is supported in $B(t)$ and

$$\int_{B(t)} p \, dx = A \int_{B(t)} \rho^\gamma \exp S \, dx \geq A(\operatorname{vol} B(t))^{1-\gamma} \left(\int_{B(t)} \rho \exp \frac{S}{\gamma} \, dx\right)^\gamma$$
$$= A(\operatorname{vol} B(t))^{1-\gamma} \left(\eta(t) + \operatorname{vol} B(t) \bar{\rho} \exp \frac{\bar{S}}{\gamma}\right)^\gamma$$

by Hölder's inequality; hence, since $\eta(t) = \eta(0) \geq 0$,

$$\int_{B(t)} p \, dx \geq \bar{p} \operatorname{vol} B(t) = \int_{B(t)} \bar{p} \, dx.$$

bigskip

c. By the Cauchy-Schwarz inequality and point (b),

$$F^2(t) \leq F'(t) \int_{B(t)} |x|^2 \rho \, dx;$$

moreover, $\int_{B(t)} |x|^2 \rho \, dx \leq (R + \sigma t)^2 \left(m(t) + \int_{B(t)} \bar{\rho} \, dx\right)$, hence, since $m(t) = m(0)$,

$$F^2(t) \leq F'(t)(R + \sigma t)^2 \frac{4\pi}{3} \left[\max \rho^0 R^3 + \bar{\rho}((R + \sigma t)^3 - R^3)\right] \leq$$
$$\leq F'(t) \frac{4\pi}{3} (R + \sigma t)^5 \max \rho^0(x)$$

since $\bar{\rho} \leq \max \rho^0$.

d. From this differential inequality, we get, for $t < \bar{T}$, $F(t) > 0$ (because $F(0) > 0$) and

$$F(0)^{-1} > F(0)^{-1} - F(t)^{-1} \geq (\alpha \sigma \max \rho^0)^{-1}(R^{-4} - (R + \sigma t)^{-4}).$$

This implies $\bar{T} < \infty$. If (3.6) is strengthened to $F(0) > CR^4, C = \alpha\sigma\max\rho^0$, we obtain the estimate

$$(3.7) \qquad \bar{T} \leq T_0 = \frac{R}{\sigma}\Big[\Big(\frac{F(0)}{F(0) - CR^4}\Big)^{\frac{1}{4}} - 1\Big].$$

$$\diamond$$

In Example (b) of the semilinear wave equation, the simple method explained there using $F(t) = \int u(x,t)dx$ is not unrealistic, because we already know, by analogy with Theorem 3.1 of Chapter I, that u itself may blow up. As a matter of fact, the method can be improved to a semisharp method, giving for instance in the case $p = 2$, a lifespan of the correct magnitude ε^{-2} (we will come back to this case in Chapter III).

For the Euler system considered in Example (c), the given proof is analogous to Example (a) and does not display realistic behavior of the solution; it is believed that only $\nabla\rho$ and ∇u do in fact blow up at time $\bar{T}$ (this has been proved only in the special case of axisymetric data in two space dimensions; see [Al6]).

4. Blowup or not? Averaging and comparison methods

In the following example of a **quasilinear wave equation**, we use a more sophisticated method than the functional method presented in section 3. One could call it the "averaging and comparison" method.

The idea is to introduce polar coordinates (r,ω) in the x-variable, and to consider the average $\bar{u}$ (in ω) of the given u. Under appropriate convexity assumptions on the nonlinear terms, this leads to a partial differential inequality (in the two variables (r,t)) on $\bar{u}$, which can be analyzed using characteristics, and so on.

Theorem 4.1. *Consider v the solution in $\mathbb{R}_t \times \mathbb{R}_x^3$ of*

$$(4.1) \qquad \partial_t^2 v - \Delta_x v = 2\partial_t v \partial_t^2 v, \quad v(x,0) = f(x), \quad \partial_t v(x,0) = g(x)$$

where $f, g \in C_0^\infty$ are supported in the ball of radius R. Assume

$$(4.2) \qquad L = (4\pi)^{-1}\int h(x)dx > 0, \quad h(x) = g(x) - g^2(x).$$

Then $\bar{T} < \infty$.

PROOF OF THEOREM 4.1.

a. For $t < \bar{T}$, v is smooth and supported in $|x| \le t + R$ (again, we refer to [Jo3] for a proof). Let $u(x,t) = \int_0^t v(x,s)ds$; u satisfies

(4.3) $\qquad u(x,0) = 0, \quad \partial_t u(x,0) = f(x), \quad \partial_t^2 u - \Delta u = w$

where $w(x,t) = (\partial_t^2 u(x,t))^2 + h(x)$.

b. To any function $p(x,t)$, we associate the function $\bar{p}$ defined by

(4.4) $\qquad\qquad \bar{p}(r,t) = (4\pi)^{-1} \int_{|\omega|=1} p(r\omega, t)\, d\omega.$

By construction, $\bar{p}$ is an even function of r, which is just the average (in x) of p on a sphere of radius r about the origin when $r > 0$.

Since, in polar coordinates, $\Delta = \partial_r^2 + \frac{2}{r}\partial_r - \frac{1}{r^2}\Delta_\omega$, we have $\overline{\Delta p} = r^{-1}\partial_r^2(r\bar{p})$. Hence $\bar{u}$ satisfies

(4.5) $\qquad\qquad \partial_t^2(r\bar{u}) - \partial_r^2(r\bar{u}) = r\bar{w}.$

This equation can be solved explicitly and we obtain

$$2r\bar{u}(r,t) = (r+t)\bar{u}(r+t,0) + (r-t)\bar{u}(r-t,0)+$$

(4.6) $\qquad\qquad + \int_{r-t}^{r+t} \rho\partial_t\bar{u}(\rho,0)d\rho + \int_{R_{r,t}} \rho\bar{w}(\rho,\tau)d\rho d\tau,$

where $R_{r,t}$ is the triangle with vertices $(r,t), (r-t,0), (r+t,0)$ in the plane.

c. We will now prove the integral inequality

(4.7) $\qquad\qquad U(s) \ge \frac{L}{s} + \frac{3}{16sR^3} \int_R^s \sigma U^2(\sigma)\, d\sigma$

for the function $U(s) = \bar{u}(s, s+R), s \ge R$. Note that because of the strong Huygens principle, U would be zero if w were zero in (4.3).

Since $\bar{w}$ is an even function of r, (4.6) gives us

$$U(s) = (2s)^{-1} \int_{T_s} \sigma\bar{w}(\sigma,\tau)\, d\sigma\, d\tau$$

where T_s is the rectangle with vertices $(R,0), (s+R, s), (s, s+R), (0,R)$.

For the part $\bar{h}(r)$ of $\bar{w}$, we have $\int_{T_s} \sigma\bar{h}(\sigma)d\sigma d\tau = 2\int_0^R \sigma^2\bar{h}(\sigma)d\sigma = 2L$, because h is zero for $|x| \ge R$.

On the other hand, the Cauchy-Schwarz inequality applied to (4.4) gives $(\bar{p})^2 \leq \overline{p^2}$; thus

$$\int_{T_s} \sigma \overline{(\partial_\tau^2 u)^2}\, d\sigma\, d\tau \geq \int_{T_s} \sigma(\partial_\tau^2 \bar{u})^2 d\sigma\, d\tau \geq \int_R^s \sigma\, d\sigma \int_{\sigma-R}^{\sigma+R} (\partial_\tau^2 \bar{u})^2\, d\tau$$

since T_s contains the parallelogram bounded by the lines $\sigma = R$, $\sigma = s$, $\tau = \sigma - R$, $\tau = \sigma + R$.

Finally, since $\partial_\tau^2(\bar{u})$ vanishes for $\tau \leq \sigma - R$, we have

$$U(s) = \int_{s-R}^{s+R} (s + R - \tau)\partial_\tau^2(\bar{u})(s,\tau)\, d\tau$$

and, by the Cauchy-Schwarz inequality,

$$U^2(s) \leq \frac{8R^3}{3} \int_{s-R}^{s+R} (\partial_\tau^2 \bar{u})^2(s,\tau)\, d\tau,$$

which is (4.7).

d. Using Gronwall's lemma, we find that $sU(s) \geq W(s)$, where W is the solution of the integral equation

$$W(s) = L + \frac{3}{16R^3} \int_R^s \sigma^{-1} W^2(\sigma)\, d\sigma.$$

In fact, $W(s) = L\left(1 - \frac{3L}{16R^3} \ell n\, \frac{s}{R}\right)^{-1}$; hence $\bar{T} \leq 2R \exp \frac{16R^3}{3L}$.   ◇

Again, as emphasized by John in his survey paper [Jo6], one expects only the second order derivatives of v to blow up at time $\bar{T}$; this has been proved only in the axisymmetric case, which we will consider in detail in Chapter IV.

Roughly speaking, one can say that the above proofs rely in an essential way on the sign and the convexity of the nonlinearity, and, in some cases, on the space dimension (because up to dimension three, the elementary solution of the wave operator is positive). In the case of the three dimensional semilinear wave equation with nonlinear terms of the opposite sign

$$\partial_t^2 u - \Delta u = -|u|^{p-1}u,$$

no blowup result is known. In fact, one can prove $\bar{T} = \infty$ for $p \leq 5$ (5 is said to be the "critical index"); it is not known whether $\bar{T} = \infty$ for all p.

Notes

The first two sections are taken from Majda [Ma] (Chapter II), where more details can be found about hyperbolic symmetric systems. For some background on symmetric systems, see for example Friedrichs [Fr] or, more recently, Rauch [Ra].

An extension to the case of strictly hyperbolic systems using a pseudodifferential symmetrizer can be found in Métivier [Me].

Specific blowup criteria have been established for the incompressible and the compressible Euler systems (see respectively Beale, Kato and Majda [BKM] and Chemin [Ch]).

There is important literature about functional methods. One can find in Levine [Le1] and Keller [Ke] a good survey of earlier references. The more recent references are given in the two survey papers by John [Jo6] and Strauss [St].

The presentation of Example (b) of Section 3 is due to Sideris [Si1], though the example itself goes back to John [Jo1] and Glassey [Gl1], [Gl2]. Again, Example (c) of Section 3 is taken from Sideris [Si3], while Section 4 is due to John [Jo6].

We avoided discussing the huge literature on global existence for semilinear wave equations. A recent survey is due to Zuily [Zu].

CHAPTER III

Semilinear Wave Equations

Introduction

In the general framework of quasilinear systems or equations, we could only define in Chapter II the lifespan of a solution. For semilinear systems or equations, it is possible to go further and define a maximal domain of existence of a given solution. The essential questions are then about the shape of this domain and the behavior of the solution near its boundary.

In the first section, we discuss the modifications of Theorems 2.2 and 2.3 of Chapter II in the semilinear cases, and formulate the corresponding blowup criteria.

The second section contains a brief discussion of the concept of maximal influence domain for smooth solutions of semilinear wave equations.

In Section 3, we extend this concept to weak solutions and prove an analogue of Theorem 2.2 of Chapter II; this approach makes it also possible to give a strong blowup criterion at the boundary of the maximal influence domain.

Finally, in a rather special situation, we obtain uniform blowup rates along the space-like boundary of the maximal influence domain, a picture (locally) very close to the examples displayed in Chapter I A.

1. Semilinear blowup criteria

Consider a system

$$(1.1) \qquad Lu = \partial_t u + \sum_{j=1}^{n} A_j \, \partial_j u + B(u) = 0$$

with the same hyperbolicity assumptions as in Chapter II, the only difference being that the coefficients A_j **do not depend** on u any more. We assume for simplicity that B is defined **everywhere**.

The short time existence Theorem 1.1 of Chapter II can then be modified in the following way.

Theorem 1.1. *Consider a semilinear system (1.1). For some integer $s > \frac{n}{2}$, assume that*

$$(1.2) \qquad\qquad u_0 \in H^s.$$

Then, for any $M \geq |u_0|_s$, there exists $T > 0$, depending only on M and B, and a unique u solution of (1.1) for $0 \leq t \leq T$ satisfying

$$(1.3) \qquad u \in C^0([0,T], H^s) \cap C^1([0,T], H^{s-1}), \quad u(x,0) = u_0(x).$$

This theorem can be proved exactly as its homologue (Theorem 2.2) of Chapter I.

It is important here to remark that, just as in Theorem 2.2 of Chapter I, u is not close to u_0; only $|u|_s$ is bounded by a constant times $|u_0|_s$.

The analogue of Theorem 2.3 is

Theorem 1.2. *Consider u a C^1 solution of (1.1) for $0 \leq t < T < \infty$ with initial value $u_0 \in H^s$, $s > \frac{n}{2}$. Assume that there exists M such that, for $0 \leq t < T$,*

$$(1.4) \qquad\qquad |u(x,t)| \leq M.$$

Then

$$(1.5) \qquad u \in C^0([0,T'], H^s) \cap C^1([0,T'], H^{s-1}), \quad T' < T$$

and

$$(1.6) \qquad\qquad |u(.,t)|_s \leq C|u_0|_s, \quad t < T,$$

where C depends only on s and M.

The proof is a simple modification of that of Theorem 2.3, using the above Theorem 1.1.

The whole "theory" of the lifespan can then be repeated as in Chapter II, and the blowup criterion corresponding to Theorem 1.2 is the following.

Blowup criterion. *Let $\bar{T}$ be the lifespan of a solution u of (1.1) with initial value $u_0 \in C_0^\infty$. Then u is not bounded for $t < \bar{T}$.*

For semilinear equations

$$(1.7) \qquad \partial_t^2 u + \sum_{i,j \geq 0} g_{ij} \partial_{ij}^2 u + F(u, \nabla u) = 0.$$

A completely analogous theory can be developed if F depends only on ∇u. In this case, the first order derivatives of u are the unknowns of the corresponding first order system, and blowup occurs only if $|\nabla u|$ is not bounded (the restriction on F comes from the fact that the energy inequality controls only ∇u). In the general case, assuming that the initial data have compact support and using finite propagation speed, one can control u by ∇u and obtain the same result.

Finally, if F depends only on u, similar considerations lead to the fact that blowup can only occur if $|u|$ in not bounded.

2. Maximal influence domain

In the remaining part of this chapter, we restrict ourselves to semi-linear wave equations

$$(2.1) \quad \partial_t^2 u - \Delta u + F(u, \nabla u) = 0, \quad u(x,0) = f(x), \quad \partial_t u(x,0) = g(x).$$

Let us denote by $C(x^0, t^0)$ the open backward cone

$$(2.2) \qquad C(x^0, t^0) = \{(x,t), t - t^0 < -|x - x^0|\}.$$

We introduce the following definition.

Definition 2.1. *An open set Ω is called an* **influence domain** *if $(x,t) \in \Omega$ implies $\bar{C}(x,t) \subset \Omega$.*

It follows from this definition that $\phi(x) = \max\{t, (x,t) \in \Omega\}$ is either identically $\pm\infty$ or else Lipschitz continuous with

$$(x,t) \in \Omega \Leftrightarrow t < \phi(x)$$

and

$$|\phi(x) - \phi(y)| \leq |x - y|.$$

We define now the maximal influence domain.

Theorem and Definition 2.2. *Consider equation (2.1) with initial data* $f, g \in C_0^\infty$. *Let* $\Omega_{\max}$ *be the union of all influence domains* Ω *containing* $\{t \leq 0\}$ *such that there is a* C^∞ *solution of (2.1) in* $\Omega \cap \{t \geq 0\}$ *with initial data* f, g. *Then* $\Omega_{\max}$ *is the unique maximal domain with this property.*

PROOF OF THEOREM 2.2.

a. We know that there exist $T > 0$ and a C^∞ solution u of (2.1) for $t < T$ with initial data f, g. Thus the class $\mathcal{S}$ of all the influence domains considered in the theorem is not empty.

b. Let (Ω_1, u^1) and (Ω_2, u^2) two domains of $\mathcal{S}$ with their corresponding solutions. If $(x^0, t^0) \in \Omega_1 \cap \Omega_2$, $t^0 > 0$, u^1 and u^2 are two C^∞ solutions of (2.1) in $\bar{C}(x^0, t^0) \cap \{t \geq 0\}$ with the same initial data. Hence $u^1(x^0, t^0) = u^2(x^0, t^0)$. We have obtained a smooth solution u in $\Omega_1 \cup \Omega_2$.

$\diamond$

By the definition of $\Omega_{\max}$, the supremum of all T such that the strip $\{t < T\}$ is contained in $\Omega_{\max}$ is just the lifespan $\bar{T}$. Since we restricted ourselves to initial data with compact supports say, in $\{|x| \leq R\}$, $\Omega_{\max}$ contains in fact the open set $\{t < \bar{T}\} \cup \{|x| > t+R\}$. Hence the hyperplane $\{t = \bar{T}\}$ intersects the complement of $\Omega_{\max}$ along a compact K; using the blowup criterion, we see that $(u, \nabla u)$ cannot remain in a compact subset of G in some neighborhood of K. In other words, blowup must occur at some boundary point of $\Omega_{\max}$ in K.

3. Maximal influence domains for weak solutions

In the preceding section, we have defined the maximal influence domain $\Omega_{\max}$ of a smooth solution u, and we have observed that $|u| + |\nabla u|$ cannot remain bounded near certain points of $\partial \Omega_{\max}$. The proof relies on local existence of smooth solutions and global uniqueness in cones.

If we consider, instead of C^∞ solutions, solutions with a limited smoothness (expressed by $u \in \mathcal{W}$ for some Banach space $\mathcal{W}$), we can define as well a maximal influence domain provided solutions in $\mathcal{W}$ exist locally and are unique in cones. It turns out that this influence domain is essentially independent of $\mathcal{W}$, if $\mathcal{W}$ is small enough. While, in Theorem

2.2 of Chapter II, we considered strips $\{0 \le t \le T\}$ and Sobolev regularity in x, we consider here analogously influence domains and C^k smoothness. This approach allows us also to analyze the behavior of the solution close to the boundary of the domain.

We will develop this point of view in the special case of the equation

$$(3.1) \qquad \partial_t^2 u - \Delta u = u^2, \quad u(x,0) = f(x), \quad \partial_t u(x,0) = g(x)$$

in three space dimensions.

Remark first that if $u \in C^2$ is a classical solution of (3.1) in an influence domain Ω, then

$$(3.2) \qquad \tilde{u} = E * \tilde{u}^2 + \tilde{u}_0.$$

Here, the tilde denotes extension by 0 for $\{t < 0\}$, the star the convolution, u_0 the free solution with data f, g, and E the usual fundamental solution of the wave equation

$$(3.3) \qquad E = (2\pi)^{-1} \delta(t^2 - |x|^2).$$

Thus formula (3.2) reads for $t \ge 0$

$$(3.4) \quad u(x,t) = u_0(x,t) + (4\pi)^{-1} \int_0^t (t-s)ds \int_{|\omega|=1} u^2(x + (t-s)\omega, s) \, d\omega.$$

We introduce now the space of weak solutions of (3.2).

Definition 3.1 . *Let Ω be an influence domain. Then*

$$(3.5) \quad \mathcal{W}(\Omega) = \Big\{ u, t < 0 \Rightarrow u(x,t) = 0, \ u \in L^2_{\mathrm{loc}}(\Omega), \ E * u^2 \in L^\infty_{\mathrm{loc}}(\Omega) \Big\}.$$

We define in the same way $\mathcal{W}(\bar{\Omega})$.

Remark that $\mathcal{W}$ is a vector space (because E is positive), and that functions in $\mathcal{W}$ need not be bounded.

The main result of this section is the following.

Theorem 3.1. *Assume $u_0 \in \mathcal{W}(\mathbb{R}^4)$. Then there exists a unique maximal influence domain $\Omega_{\max}$ containing $\{t \le 0\}$ and a unique solution $u \in \mathcal{W}(\Omega_{\max})$ of (3.2). If $(x^0, t^0) \in \partial\Omega_{\max}$ and*

$$(3.6) \qquad \bar{C}(x^0, t^0) \subset (x^0, t^0) \cup \Omega_{\max},$$

then $E * u^2$ is unbounded in $\Omega_{\max} \cap V$ for any neighborhood V of (x^0, t^0). Moreover, if

$$u_0 \in C^k(\Omega_{\max} \cap \{t \geq 0\}),$$

then

$$u \in C^k(\Omega_{\max} \cap \{t \geq 0\}).$$

The proof will be divided into three steps.

Step 1: Local existence and uniqueness in a cone

a. We prove first a uniqueness result.

Uniqueness lemma.

(i) Let Ω be an influence domain, $u_1 \in L^\infty(\Omega)$, $u_1 = 0$ for $t < 0$ and $u_2 = E * u_1$. Set

$$m_i(t) = \max |u_i(., t)|, \quad i = 1, 2,$$

where max means the essential supremum in x for $(x, t) \in \Omega$. Then

$$m_2(t) \leq \int_0^t (t - s) m_1(s) ds.$$

(ii) Assume $v, h \in L^1[0, T]$ and

$$v(t) \leq C \int_0^t (t - s) v(s) ds + h(t).$$

Then

(3.7) $\qquad v(t) \leq C^{\frac{1}{2}} \int_0^t \sin h(C^{\frac{1}{2}}(t - s)) h(s) ds + h(t).$

(iii) If $u_i \in \mathcal{W}(\bar{C}(x^0, t^0))$, $i = 1, 2$, are two solutions of an equation

$$u = E * u^2 + v$$

in a cone $\bar{C}(x^0, t^0)$, then $u_1 = u_2$.

PROOF OF THE UNIQUENESS LEMMA.

(i) This point is an immediate consequence of (3.4).

(ii) Let $H(s) = 1$ for $s \geq 0$ and $H(s) = 0$ for $s < 0$. Set $v_0(s) = v(s)H(s), h_0(s) = h(s)H(s), E_0(s) = sH(s)$ and $w = E_0 * v_0$. Then

$$w'' = v_0 \leq Cw + h_0.$$

Since the fundamental solution

$$E_C(t) = C^{-\frac{1}{2}} H(t) \sin h(C^{\frac{1}{2}}t)$$

of $\frac{d^2}{dt^2} - C$ is positive, it follows that

$$w(t) \leq E_C * h_0(t),$$

which gives (3.7).

(iii) We have

$$u_1 - u_2 = E * (u_1 - u_2)(u_1 + u_2),$$

hence

$$|u_1 - u_2|^2 \leq 2[E * (u_1^2 + u_2^2)][E * |u_1 - u_2|^2]$$

because of the Cauchy-Schwarz's inequality which gives

$$(E * fg)^2 \leq (E * f^2)(E * g^2).$$

Since the $E*u_i^2$ are bounded, the uniqueness lemma (i) can be applied to the function $|u_1 - u_2|^2$, and (ii) of the same lemma implies then $u_1 = u_2$.   ◇

b. The local existence result is the following.

Local existence lemma. *Let*

$$v \in \mathcal{W}(\bar{C}(x^0, t^0)), \quad t^0 > 0, \quad E * v^2 \leq C_0$$

and assume $8C_0(t^0)^2 \leq 1$. *Then the equation*

(3.8) $u = E * u^2 + v$

has a unique solution $u \in \mathcal{W}(\bar{C}(x^0, t^0))$ *with* $E * u^2 \leq 4C_0$.

PROOF OF THE LOCAL EXISTENCE LEMMA. We use a fix point argument, setting $u^0 = v$, $u^{k+1} = E * (u^k)^2 + v$. Assume first $v \geq 0$. Then $u^1 \geq u^0 \geq 0$, and $u^k \geq u^{k-1} \geq 0$, $k \geq 1$, implies

$$u^k = E * (u^{k-1})^2 + v \leq E * (u^k)^2 + v = u^{k+1}.$$

Hence the sequence u^k is increasing, and

$$E * (u^{k+1})^2 \leq 2E * (E * (u^k)^2)^2 + 2E * v^2.$$

The induction hypothesis $E * (u^k)^2 \leq 4C_0$ implies, by the uniqueness lemma (i) and our assumption $8C_0(t^0)^2 \leq 1$

$$E * (u^{k+1})^2 \leq 16C_0^2(t^0)^2 + 2C_0 \leq 4C_0.$$

Hence $u^k - v$ is bounded, so $u = \lim u^k$ exists and satisfies (3.8) .

If v may be negative, let w be the solution of $w = E * w^2 + |v|$, and define also w^k as before. By induction, one obtains

$$|u^k| \leq w^k, \quad |u^{k+1} - u^k| \leq w^{k+1} - w^k.$$

Hence the convergence of w^k implies the convergence of u^k. $\diamondsuit$

Step 2: Existence of $\Omega_{\max}$ and behavior at the boundary

a. The proof of the existence of $\Omega_{\max}$ is identical to the proof of Theorem 2.2, using the lemmas of Step 1.

b. Let $(x^0, t^0) \in \partial\Omega_{\max}$ satisfying (3.6) and assume that there exists a neighborhood V of (x^0, t^0) such that $E * u^2$ is bounded in $V \cap \Omega_{\max}$. For ε_1 small enough,

$$\bar{C}(x^0, t^0 + \varepsilon) \cap \Omega_{\max}^c \subset V, \quad \varepsilon \leq \varepsilon_1,$$

the letter c denoting the complement. Let

$$v = (1 - \chi)(E * (\chi u^2) + u_0),$$

where χ is the characteristic function of $\Omega_{\max}$. We have $E * v^2 \leq C'$ in $\bar{C}(x^0, t^0 + \varepsilon_1)$. Since

$$\bar{C}(x^0, t^0 + \varepsilon) \cap \Omega_{\max}^c \subset \{t > t^0 - \delta\}$$

for $\varepsilon \leq \varepsilon_2(\delta)$, we can solve $w = E * w^2 + v$ in $\bar{C}(x^0, t^0 + \varepsilon)$ if

$$8C'(\varepsilon + \delta)^2 \leq 1, \quad \varepsilon \leq \varepsilon_1, \quad \varepsilon \leq \varepsilon_2(\delta),$$

according to the local existence lemma. Then w is supported in $\Omega_{\max}^c$ and $\bar{u} = w + \chi u$ satisfies

$$\bar{u} = E * w^2 + v + \chi(E * (\chi u^2) + u_0) = E * \bar{u}^2 + u_0$$

in $C(x^0, t^0 + \varepsilon) \cup \Omega_{\max}$. We have thus obtained an extension of u to a larger set than $\Omega_{\max}$, a contradiction.

Step 3: Additional smoothness of the solution

Choose cones C and C' such that

$$\bar{C} \subset C' \subset \bar{C}' \subset \Omega_{\max}.$$

A. Assume u_0 continuous in $\Omega_{\max} \cap \{t \geq 0\}$. Then $|u_0| \leq M$, $E * u^2 \leq M$ in C', hence $|u| \leq 2M$ in C'.

a. For $t^0 > 0$ small enough, we can use the proof of the local existence and uniqueness lemma in each cone $\bar{C}(x^0, t^0) \subset C'$ to see that u coincides almost everywhere in $\bar{C}(x^0, t^0)$ with the everywhere limit $\bar{u}$ of the continuous u^k. Since, as a consequence of the uniqueness lemma (i),

$$|u^k - u_0| \leq w^k - |u_0| \leq At^2,$$

the continuity of u_0 at a point $(x^0, 0)$ implies that of $\bar{u}$ at the same point.

b. Let now $(x^0, t^0) \in C$, $t^0 > 0$. We denote by $T_h u(.) = u(h + .)$ the translation of u with $h = (h_0, \ldots, h_3)$, $h_0 < 0$, $h + C \subset C'$. Since u_0 is continuous in $C' \cap \{t \geq 0\}$, we have

(3.9) $$|T_h u_0(x, t) - u_0(x, t)| \leq \varepsilon + M \chi_h(t)$$

in C for $|h| \leq \delta_1(\varepsilon)$, where $\chi_h(t) = 1$ for $0 \leq t \leq -h_0$ and 0 otherwise. By the properties of convolution, we have

$$T_h u - u = E * ((T_h u + u)(T_h u - u)) + T_h u_0 - u_0,$$

hence

$$w_h \leq 4CE * w_h + \varepsilon + M \chi_h(t), w_h = |T_h u - u|.$$

With $m(t) = \max_C w_h(., t)$, the uniqueness lemma gives

$$m(t) \leq 4M \int_0^t (t - s) m(s) ds + \varepsilon + M \chi_h(t),$$

and thus

$$m(t) \leq M_1(\varepsilon + |h|) + M \chi_h(t).$$

For (x, t) in a fixed neighborhood of (x^0, t^0), we will have in particular $|T_h u - u|(x, t) \leq \varepsilon$ if $|h| \leq \delta_2(\varepsilon)$. Consider now standard regularizations

u^η of u, for which $u^\eta \to u$ almost everywhere near (x^0, t^0) (because we already know $u \in L^2$). We have

$$|h| \leq \delta_2(\varepsilon) \Rightarrow |T_h u^\eta - u^\eta| \leq \varepsilon,$$

hence the u^η converge everywhere to a continuous function $\bar{u}$; $\bar{u}$ is equal to u almost everywhere close to (x^0, t^0).

We have proved that u is continuous in $C \cap \{t \geq 0\}$.

B. Assume now $u_0 \in C^k, k \geq 1$.

a. Since u_0 is Lipschitz continuous, (3.9) holds with $\varepsilon = M_2|h|$, and it follows that u is Lipschitz continuous for $t \geq 0$.

b. We have $\partial_{x_i} u^2 = 2uw$ with $w = \partial_{x_i} u$, hence

$$w = 2E * (uw) + w_0$$

where $w_0 = \partial_{x_i} u_0 \in C^{k-1}$ for $t \geq 0$. We obtain the same equation for $w = \partial_t u$ with now

$$w_0 = \partial_t u_0 + E * \mu, \quad \mu = \delta(t) \otimes f^2(x).$$

Since $E * \mu$ is just $Ct \int_{|\omega|=1} f^2(x + t\omega)d\omega$, we have $w_0 \in C^{k-1}$ for $t \geq 0$ also in this case.

Now

$$T_h w - w = 2E * (T_h w T_h u - wu) + T_h w_0 - w_0.$$

Using the bound for $E*|T_h u - u|$ obtained in A., we get for $v_h = |T_h w - w|$ the estimate

$$v_h \leq 2ME * v_h + M_2(|h| + M\chi_h(t)) + |T_h w_0 - w_0|.$$

We may now proceed as before to conclude that w is continuous and Lipschitz continuous if $k \geq 2$. Repeating this process yields $u \in C^k$. $\diamond$

Some remarks are in order after this proof:

(i) If we take initial data $f \in C_0^{k+1}$, $g \in C_0^k$, we will have $u_0 \in C^k$ in $\Omega_{\max}$. In particular, if $k = +\infty$, we are in the simple situation of Theorem 2.2 above.

(ii) The blowup criterion "$E * u^2$ is unbounded" of Theorem 3.1 is an improved version of the usual "u is bounded" in the spirit of the criterion (1.2) obtained in Chapter I for ODEs.

4. Blowup rates at the boundary of the maximal influence domain

We will now consider a very special situation where it is possible to obtain a rather complete picture of what happens at the boundary of $\Omega_{\max}$.

Let $f(x)$, $g(x)$ be defined and smooth for $|x| \leq R + T$, where R and T are given positive constants. Consider the solution u of

(4.1) $\partial_t^2 u - \Delta u = F(u)$, $F(u) = |u|^p$, $p > 1$,

$$u(x, 0) = f(x), \ \partial_t u(x, 0) = g(x), \ x \in \mathbb{R}^3$$

in (a subdomain of) $K_{R,T} = \underset{|x| \leq R}{\cup} (\bar{C}(x, T) \cap (t \geq 0))$.

We make first the two following assumptions on the initial data.

Assumptions I.

(4.2)
$$f(X) + tg(X) > t|\nabla f(X)|, \ \ X = x + t\omega, \ |x| \leq R, \ 0 \leq t \leq T, \ |\omega| \leq 1.$$

For some $\eta > 0$,

$$g(X) - (1 + \eta)|\nabla f(X)| + t(\Delta f(X) + F(f(X)) - (1 + \eta)|\nabla g(X)|) >$$
(4.3) $$> t|\nabla g(X)| + (1 + \eta)t|\nabla^2 f(X)|.$$

These strange looking assumptions are related to the explicit representation of solutions of the linear wave equation in three dimensions: the solution u of

$$\partial_t^2 u - \Delta u = h, \ \ u(x, 0) = f(x), \ \partial_t u(x, 0) = g(x)$$

is given by

$$4\pi u(x, t) = \int_{|\omega|=1} \left\{ f(x + t\omega) + t\Big[(\omega\nabla)f(x + t\omega) + g(x + t\omega)\Big] \right\} d\omega +$$
(4.4) $$+ \int_0^t (t - s)ds \int_{|\omega|=1} h(x + (t - s)\omega, s)d\omega.$$

The assumptions (4.2), (4.3) have the following consequences: consider the solution u_1 of

$$\partial_t^2 u_1 - \Delta u_1 = 0, \ \ u_1(x, 0) = f(x), \ \partial_t u_1(x, 0) = g(x).$$

Then

(i′) (4.2) implies $u_1 \geq 0$ in $K_{R,T}$.

(ii′) (4.3) implies $\partial_t u_1 \geq (1 + \eta)|\nabla u_1|$ in $K_{R,T}$.

More generally, conditions (4.2) and (4.3) will always ensure the positivity of various solutions of the wave equation appearing in the proof.

Finally, to actually see some blowup in the cylinder $|x| \leq R$, $t \leq T$, we impose the rough following size assumption on the initial data.

Assumption II. *Choose constants γ_1 and γ_2 such that the solution w of*

$$w''(t) = F(w), \quad w(0) = \gamma_1, \quad w'(0) = \gamma_2, \quad t > 0$$

blows up at a time T_1, $0 < T_1 < T$ and moreover $\gamma_1 > T_1(\gamma_1 - \gamma_2)$. We assume then

(4.5) $$f \geq 2\gamma_1, \quad g \geq 2\gamma_2, \quad |\nabla f| < \gamma_1.$$

The picture obtained is described in the following theorem.

Theorem 4.1. *Suppose (4.2), (4.3) and (4.5). Then there exists a classical solution u of (4.1) in $\Omega = \{(x,t), |x| \leq R + T, 0 \leq t < \phi(x)\}$, satisfying*

(a)

    (i) $0 < \phi(x) < T$ for $|x| \leq R$, $0 < \phi(x) \leq R + T - |x|$ for $R \leq |x| \leq R + T$,

    (ii) ϕ *is Lipschitz continuous with constant 1 in $|x| \leq R + T$, with constant at most $(1 + \eta)^{-1}$ for $|x| \leq R$.*

(b)

    (i) $u(x,t) \rightarrow +\infty$ when $t \rightarrow \phi(x)$, $|x| \leq R$,

    (ii) *Let $d = d(x,t)$ denote the distance from (x,t) to $\partial\Omega$. Then there exist $\delta > 0$ and positive c, C such that we have, for $|x| \leq R$, $d \leq \delta$, with $q = \frac{2}{p-1}$,*

(4.6) $$c \leq u d^q \leq C, \ c \leq d^{q+1}\partial_t u \leq C, \ d^{q+1}|\nabla u| \leq C, \ c \leq d^{q+2}\partial_t^2 u \leq C.$$

PROOF OF THEOREM 4.1.

A. Construction of u, ϕ, Ω.

a. As usual, the solution u is obtained by a fix point argument. Define $u_0 = 0$ and u_{n+1} by

$$(4.7) \quad (\partial_t^2 - \Delta)u_{n+1} = F(u_n), \quad u_{n+1}(x,0) = f(x), \quad \partial_t u_{n+1}(x,0) = g(x).$$

Then $0 \le u_n(x,t) \le u_{n+1}(x,t)$ in $K_{R,T}$.

In fact, $u_1 \ge 0$ by condition (4.2), and, by the monotony of F and (4.4), $u_n - u_{n-1} \ge 0$ implies $u_{n+1} - u_n \ge 0$.

b. We prove now $\partial_t u_n \ge (1+\eta)|\nabla u_n|$ in $K_{R,T}$.

Let e be a unit vector in $\mathbb{R}^3$ and set $J_n = \partial_t u_n + (1+\eta)e\nabla u_n$. We have

$$(\partial_t^2 - \Delta)J_{n+1} = F'(u_n)J_n, \quad J_{n+1}(x,0) = [g + (1+\eta)e\nabla f](x),$$

$$(4.8) \qquad\qquad \partial_t J_{n+1}(x,0) = [\Delta f + F(f) + (1+\eta)e\nabla g](x).$$

Clearly, $J_0 = 0$; suppose $J_n \ge 0$: then $F'(u_n)J_n \ge 0$, and condition (4.3) and (4.4) imply $J_{n+1} \ge 0$.

c. Let $u(x,t) = \lim u_n(x,t)$ (possibly $+\infty$). Since u is increasing in t, there is a function $\phi(x)$, defined in $|x| \le R + T$, such that

(i) $0 < \phi(x)$, $\phi(x) \le T$ if $|x| \le R$ and $\phi(x) \le R + T - |x|$ if $R \le |x| \le R + T$,

(ii) $u(x,t) < +\infty$ if $t < \phi(x)$, $u(x,t) = +\infty$ if $t > \phi(x)$.

Remark that at this stage, we have not proved yet any blowup in $K_{R,T}$. This will be done in (f).

d. Consider the set $\Omega = \{(x,t) \in K_{R,T}, 0 \le t < \phi(x)\}$: we prove that Ω is an influence domain (strictly speaking, we should add to Ω the half-space $\{t \le 0\}$). Let $m = (x,t)$ and $m' = (x',t')$ be two points in $K_{R,T}$ with $t' < t$, $t - t' \ge (1+\eta)^{-1}|x - x'|$; since

$$u_n(x',t') = u_n(x,t) + (t'-t)\left[\int_0^1 \left(\partial_t u_n(p) + \frac{x'-x}{t'-t}\nabla u_n(p) \right) ds \right],$$

$$p = m + s(m' - m),$$

the result of (b) implies $u_n(x',t') \le u_n(x,t)$. Hence $m' \in \Omega$ if $m \in \Omega$ and in particular, Ω is an influence domain.

We can also conclude from this that ϕ is Lipschitz with constant at most $(1 + \eta)^{-1}$ for $|x| \leq R$.

e. Using (d), we see that any point in Ω belongs to some closed truncated cone K' contained in Ω; we prove now $u \in C^2(K')$. First, $u_n \leq C_0$. Let $w_n = \partial_t u_n$; we have

$$(4.9) \quad (\partial_t^2 - \Delta)w_n = F'(u_{n-1})w_{n-1}, \quad w_n(x, 0) = g(x),$$
$$\partial_t w_n(x, 0) = (\Delta f + F(f))(x).$$

We shall compare w_n with the function $W = M \exp Bt$ which satisfies

$$(4.10) \qquad (\partial_t^2 - \Delta)W = B^2 W, \quad W(x, 0) = M, \quad \partial_t W(x, 0) = MB.$$

Let us prove $W \geq w_n$ in K' for all n if M and B are chosen big enough.

We have $W \geq w_0 = 0$, and suppose $W \geq w_{n-1}$; the function $z = W - w_n$ satisfies, if $B^2 \geq F'(C_0)$,

$$(\partial_t^2 - \Delta)z = B^2 W - F'(u_n)w_{n-1} \geq (B^2 - F'(u_n))W \geq 0,$$

$$(4.11) \qquad z(x, 0) = M - g(x), \quad \partial_t z(x, 0) = MB - (\Delta f + F(f))(x).$$

If $M > |g|$, $MB > |\nabla g| + |\Delta f + F(f)|$, we conclude from (4.11) and the representation formula that $z \geq 0$.

Similarly, we prove $-w_n \leq W$; hence $|\partial_t u_n| \leq C$, and also $|\nabla u_n| \leq C$ because of (b).

To bound the second derivatives of u, set $w_n = D^2 u_n$, where D^2 means some second order derivative. We have now

$$(4.12) \qquad (\partial_t^2 - \Delta)w_n = F'(u_{n-1})w_{n-1} + F''(u_{n-1})Du_{n-1}Du_{n-1},$$

and the last term in (4.12) is bounded. We can now proceed exactly as before, comparing w_n with $W = M \exp Bt$ for an appropriate choice of M, B independent of n.

We thus obtain $|D^2 u_n| \leq C$, and so on for higher order derivatives.

f. Finally, we prove that $\phi(x) < T$ for $|x| \leq R$. This will be done by comparing u with the function w in (4.5). Set $\Omega_1 = \Omega \cap (t < T_1)$. The function $u - w$ is defined in Ω_1 (which is an influence domain) and satisfies

$$(\partial_t^2 - \Delta)(u - w) = F(u) - F(w), \quad (u - w)(x, 0) = f(x) - \gamma_1,$$
$$(4.13) \qquad\qquad\qquad \partial_t(u - w)(x, 0) = g(x) - \gamma_2.$$

For t small enough, $u > w$. We claim that $u > w$ in Ω_1. Otherwise, there would be a smallest value t_0 such that $u = w$ at some $(x_0, t_0) \in \Omega_1$; but $u \geq w$ in the backward cone $\bar{C}(x_0, t_0)$, hence (4.13) and the representation formula imply, because of (4.3), $u > w$ at (x_0, t_0), a contradiction.

Since, by construction, w does blow up at time T_1, we get $\phi(x) \leq T_1$.

B. Blowup rates

a. Let $J_n = \partial_t^2 u_n - F(u_{n-1}) + M \partial_t u_n$, $M > 0$. We verify easily that $(\partial_t^2 - \Delta) J_1 = 0$ and that J_{n+1} satisfies, for $n \geq 1$,

$$(\partial_t^2 - \Delta) J_{n+1} = F'(u_n) J_n + F''(u_n) |\nabla u_n|^2,$$

(4.14)
$$J_{n+1}(x, 0) = (f_1 + Mg)(x), \quad \partial_t J_{n+1}(x, 0) = [g_1 + M(\Delta f + F(f))](x),$$

where f_1, g_1 are functions independent of n and M.

The condition required on the traces of J_n to obtain positivity reduces, for M large, to

$$g(X) + t[\Delta f(X) + F(f(X))] > t|\nabla g(X)|, \quad X = x + t\omega$$

which is a consequence of (4.3). Hence $J_1 \geq 0$, and, recursively, $J_n \geq 0$.

b. Similarly, set $H_n = \frac{\eta}{1+\eta} \partial_t^2 u_n - F(u_{n-1}) - M \partial_t u_n$. We obtain

$$(\partial_t^2 - \Delta) H_{n+1} = F'(u_n) H_n - \left(\frac{1}{1+\eta} (\partial_t u_n)^2 - |\nabla u_n|^2\right) F''(u_n) \leq F'(u_n) H_n.$$

Moreover, the positivity condition (4.2) holds for the initial data of $-H_n$; therefore, as in (a), we obtain $H_n \leq 0$.

c. From (b), we deduce $\partial_t^2 w \leq C w^p + M \partial_t w$, $w = u_n$.

Multiplying by $\partial_t w \exp -2Mt$ and integrating, we get

$$\partial_t w \leq [C w^{p+1} + C_1]^{\frac{1}{2}}.$$

It follows that

(4.15)
$$\int_{u_n(x_0, \tau)}^{u_n(x_0, t)} [C s^{p+1} + C_1]^{-\frac{1}{2}} ds \leq t - \tau.$$

Taking $t = \phi(x_0) + \varepsilon$ and letting $n \to \infty$, we get

$$\int_{u(x_0, \tau)}^{\infty} [C s^{p+1} + C_1]^{-\frac{1}{2}} ds \leq t_0 + \varepsilon - \tau.$$

Letting $\varepsilon \to 0$ and evaluating the integral, we obtain, for some $c > 0$,

$$u(x_0, \tau) \geq \frac{c}{(t_0 - \tau)^q}, \quad t_0 = \phi(x_0), \quad q = \frac{2}{p-1}.$$

Note that an upper bound for $\partial_t^2 w$ does not imply blowup; what we prove here is that **if** blowup occurs, it is strong enough.

d. From (b), we also obtain $\partial_t^2 u \leq F(u) + M\partial_t u$, and exactly as in (c), we get $(\partial_t u)^2 \leq C u^{p+1} + C_1$ and finally $\partial_t^2 u \leq C u^p$.

e. Using now (a) and proceeding as in (c), (d), we get $\partial_t^2 u \geq c_0 u^p - C_1, c_0 > 0$ and $\partial_t^2 u \geq c u^p$ close to $\partial\Omega$ ($c > 0$). Multiplying this last inequality by $\partial_t u$ and integrating, we find $(\partial_t u)^2 \geq c u^{p+1}$, which yields (as in (c))

$$u(x_0, \tau) \leq \frac{c}{(t_0 - \tau)^q}, \quad t_0 = \phi(x_0), \quad q = \frac{2}{p-1}.$$

$\diamondsuit$

We see that in this special situation, where monotony and positivity properties play an important role, we have a blowup behavior very close to the examples displayed in Chapter I A: the blowup surface $\partial\Omega$ is time-like, and the solution behaves uniformly, close to the surface, like an appropriate power of the distance.

The question that arises naturally in this context is: Does the solution u coincide, locally near points of $\partial\Omega$, with solutions of the type actually constructed by Kichenassamy and Littman?

To answer this question, one should first notice that Cafarelli and Friedman actually prove that the function ϕ in Theorem 4.1 is differentiable. Then, one should of course develop a differentiable version of Theorem 3.1 of Chapter I A.

5. An example of a sharp estimate of the lifespan

It seems almost impossible to give an account of the numerous papers devoted to the blowup of semilinear equations (see Notes). One of the best understood cases is again the equation

$$(5.1) \qquad \partial_t^2 u - \Delta u = |u|^p, \quad u(x,0) = f(x), \quad \partial_t u(x,0) = g(x).$$

We will only quote the results of [Lin], since a detailed proof would exceed the scope of this book.

Theorem 5.1. *Let $f \in C_0^{k+1}(\mathbb{R}^3)$, $g \in C_0^k(\mathbb{R}^3)$, $k \geq 1$, be supported in $|x| \leq M$. Denote by Ω_ε the maximal influence domain for which the Cauchy problem*

$$(5.2) \qquad \partial_t^2 u - \Delta u = u^2, \quad u(x, 2M) = \varepsilon f(x), \quad \partial_t u(x, 2M) = \varepsilon g(x),$$

has a solution $u \in C^k(\Omega_\varepsilon \cap (t \geq 2M))$. Then, when $\varepsilon \to 0$,

(i) $\varepsilon^2 \Omega_\varepsilon \to \Omega$.

(ii) *If we set $w = u - \varepsilon u_0$, we have*

$$\varepsilon^{-4} w(\varepsilon^{-2}x, \varepsilon^{-2}t) \to v(x, t)$$

in $L^2_{\text{loc}}(\Omega)$ and in $C(\Omega \cap (t > |x|))$.

(iii) *If we denote by $\bar{T}_\varepsilon$ the lifespan of u, then*

$$|\varepsilon^2 \bar{T}_\varepsilon - T| \leq C\varepsilon |\ln\varepsilon|$$

for ε sufficiently small.

We must of course define the asymptotic objects Ω, u_0, v and T appearing in Theorem 5.1.

    a) u_0 is the C^k solution of

$$(\partial_t^2 - \Delta)u_0 = 0, \quad u_0(x, 2M) = f(x), \quad \partial_t u_0(x, 2M) = g(x).$$

    b) The free solution u_0 has the asymptotic behavior

$$r u_0(r\omega, r + \rho) \to F_0(\rho, \omega) \quad r = |x|, \quad x = r\omega, \quad r \to \infty.$$

The function F_0 is called the Friedlander radiation field, and is given by

$$F_0(\rho, \omega) = (4\pi)^{-1}[R(\omega, 2M - \rho; g) - \partial_s R(\omega, 2M - \rho; f)]$$

where

$$R(\omega, s; h) = \int \delta(s - \omega y)h(y)\,dy$$

denotes the Radon transform of h.

    c) If μ is the measure

$$\mu = C(\omega)r^{-2}\delta(t - r), \quad C(\omega) = \int F_0^2(\rho, \omega)\,d\rho,$$

we consider the equation

(5.3)
$$\partial_t^2 v - \Delta v = v^2 + \mu.$$

A solution will mean a function v such that

$$v = E * v^2 + E * \mu, \quad v|_{t<0} = 0, \quad v \in L^2_{\text{loc}}, \quad E * v^2 \in L^\infty_{\text{loc}},$$

where E is the usual fundamental solution of the wave equation. Then Ω is the maximal influence domain for which (5.3) has a solution v.

d) Finally, T is the lifespan of v, that is

$$T = \inf\{t, (x, t) \in \partial\Omega\}.$$

The proof of Theorem 5.1 relies on a scaling argument (as indicated in point (ii) of the theorem), combined with comparison techniques which make use of the positivity of the fundamental solution E. We will come back to such scalings in Chapter V, when dealing with the case of small data.

Note that, in addition to a sharp estimate for $\bar{T}_\varepsilon$, an approximate shape of the existence domain Ω_ε is obtained. In particular, the first blowup takes place very **far** from the boundary of the light cone. We will see that this is in contrast with the situation for quasilinear hyperbolic wave equations, where blowup seems to take place at a finite distance from the boundary of the light cone.

Notes

The version of the blowup criterion adapted to semilinear cases is well known, though we could not find a precise reference in the litterature. For closely related discussions, we refer to Gérard and Rauch [GR].

The developments in Section 2 and 3 concerning the maximal influence domain and the corresponding blowup criteria are taken from Linblad [Lin].

Section 4 is entirely due to Cafarelli and Friedman [CF1], who have also given a one dimensional result with less restrictive assumptions [CF2].

Finally, Section 5, dealing with the case of small data, is again taken from Linblad [Lin] and seems to be the only case where the first term in an asymptotic expansion of $\bar{T}_\varepsilon$ has been obtained.

The numerous papers about semilinear wave equations quoted in the bibliography generally use "functional" or "comparison and averaging" methods to get upper bounds of the lifespan. Most of them deal only with the lifespan and discuss neither the shape of the maximal influence domain nor the mechanism of the blowup. Sharp estimates seem to be obtained only in one space dimension or in rotationally invariant cases.

CHAPTER IV

Quasilinear Systems in One Space Dimension

Introduction

We will consider here quasilinear $N \times N$ systems of the form

$$(1.1) \qquad \partial_t u + A(u)\,\partial_x u = 0,$$

or rotationally invariant wave equations of the form

$$(1.2) \qquad \partial_t^2 u - c^2(\partial_t u)\Delta u = 0.$$

Even for the simple case of Burgers' equation ($N = 1$) already considered in Chapters I and II, it is not possible to define a maximal influence domain as we have done for semilinear equations in Chapter III. This is due to the fact that solutions u_i defined in influence domains D_i need not coincide in $D_1 \cap D_2$, as is easily seen by considering various extensions of an "exterior cusp solution" defined in I 3.

For 2×2 systems, we can use the Riemann invariants to make the system diagonal and show a decoupling of the two modes in finite time, thus reducing the problem to the well known scalar case. This allows us to prove finite time blowup for systems with at least one nonlinearly degenerate eigenvalue.

As to the mechanism of the blowup, it can be shown that in general it is a geometric blowup (in the sense of Chapter I), even when the explosion takes place before the decoupling of the two modes.

These aspects are discussed in the first three sections.

For general systems, only the case of small data is well understood; approximate decoupling occurs then before blowup, and the behavior of the solution can be partially described up to the time $\bar{T}$ (the unknown lifespan). The proof of this striking result makes an essential use of certain boundedness properties of first order derivatives of the solution in L^1 norm (see Section 2). It is then rather easy to obtain upper bounds and lower bounds for the lifespan, by considering the system as an ODE along characteristics and using elementary results on blowup or existence for ODEs.

Finally, we include in this chapter the case of rotationally invariant wave equations with small data, since the proof of the upper bound for the lifespan is essentially the same as for systems in one space variable.

1. The scalar case

Though we have already discussed this case in Chapters IB3 and II2, we collect here simple facts which are essential to the understanding of the cases $N \geq 2$.

a. Lifespan

For the equation (1.1) with $N = 1$, we assume that A is genuinely nonlinear in the sense that $A'(0) \neq 0$ (see Chapter I, B 2.3). If the smooth initial value u_0 is compactly supported, the lifespan is then given by

$$(1.3) \qquad\qquad (\bar{T})^{-1} = \max - A'(u_0)u_0'.$$

On each characteristic line

$$x = X_0 + A(u_0(X_0))s, \quad t = s,$$

the derivative $q(s) = \partial_x u$ satisfies the equation

$$q' + A'(u_0(X_0))q^2 = 0.$$

If the maximum in (1.3) is attained at X_0, q blows up at time $\bar{T}$ like $C(\bar{T} - s)^{-1}$.

Note that along any other curve

$$x - x_0 = \lambda(t - \bar{T}), \quad x_0 = X_0 + A(u_0(X_0))\bar{T}, \quad \lambda \neq A(u_0(X_0))$$

reaching the blowup point, the derivative $\partial_x u$ has an integrable singularity because

$$\partial_t u + \lambda \partial_x u = (\lambda - A(u))\partial_x u.$$

b. L^1-boundedness

For $t < \bar{T}$, the solution $u(x, t)$ of (1.1) is given by the solution of the blowup system

$$\partial_T \phi = A(v), \quad \partial_T v = 0, \quad \phi(X, 0) = X, \quad v(X, 0) = u_0(X)$$

through the change of variables

$$x = \phi(X, T), \quad t = T.$$

Hence, for fixed t, $dx = \partial_X \phi \, dX$, and

(1.4) $$\int |\partial_x u| dx = \int |v'(X)| dX = \int |u_0'(X)| dX.$$

In other words, $\partial_x u$ does not blow up in the L^1 norm.

2. Riemann invariants, simple waves, L^1-boundedness

For a general system (1.1), we denote by $\lambda_1(u) < \ldots < \lambda_N(u)$ the real and distinct eigenvalues of $A(u)$, with corresponding left and right eigenvectors $\ell_j(u), r_j(u)$.

Integral curves of the field $L_j = \partial_t + \lambda_j(u(x, t))\partial_x$ are called j−characteristics.

Remark that if we make the change of variables $u = \Phi(U)$, the system becomes

$$\partial_t U + \Phi'^{-1}(U) A(\Phi(U)) \Phi'(U) \partial_x U = 0.$$

Thus the eigenvalues remain the same while the right eigenvectors are changed to $\Phi'^{-1} r_j(\Phi(U))$. In other words, r_j is a well defined vector field on the u-manifold.

a. Riemann invariants

For each j, $N - 1$ independent functions R_k^j satisfying

$$r_j(u) \partial_u R_k^j(u) = 0, \quad k = 1, \ldots, N - 1,$$

are called Riemann invariants.

In the case $N = 2$, we set simply $R_1^1 = w_2$, $R_1^2 = w_1$; since ∇w_j is colinear to ℓ_j, we obtain by multiplying the system on the left,

$$^t\nabla w_j\, \partial_t u + \lambda_j(u)\,^t\nabla w_j\, \partial_x u = 0,$$

that is, for smooth u,

$$\partial_t(w_j(u)) + \lambda_j(u)\partial_x(w_j(u)) = 0.$$

In the sequence, we will always assume that the application

$$u \mapsto (w_1(u),\ w_2(u))$$

is a diffeomorphism of the domain $D \subset \mathbb{R}^2$ we are interested in onto its image.

We will then rewrite the system (1.1) as

$$(2.1) \qquad \partial_t w + \Lambda(w)\partial_x w = 0, \quad \Lambda(w) = \begin{pmatrix} \lambda_1 & 0 \\ 0 & \lambda_2 \end{pmatrix},$$

keeping abusively the same notation λ_j for the eigenvalues expressed in terms of the w_k.

General systems cannot be made diagonal by an appropriate change of unknowns, but we still can find new unknowns $w(u)$ such that the j−axis is the integral curve of $r_j(w)$ through the origin.

b. Simples waves

Let us comment a little about the concept of a simple wave introduced in IB3. These waves are solutions of the form

$$u(x,t) = v(\zeta(x,t)), \quad \zeta(x,t) \in \mathbb{R}.$$

This is equivalent to saying that $v(\zeta)$ is an integral curve of some r_j while the function $\zeta(x,t)$ satisfies the scalar equation

$$(2.2) \qquad \partial_t\zeta + \lambda_j(v(\zeta))\,\partial_x\zeta = 0.$$

We will call such a solution a j−simple wave.

Note that for a j−simple wave, we have

$$(2.3) \qquad \partial_x u(x,t) = r_j(u)\partial_x\zeta(x,t),$$

and

$$(2.4) \qquad (\partial_x\zeta)(dx - \lambda_j(u)dt) = d\zeta.$$

c. L^1-boundedness

The L^1-boundedness already observed for scalar equations (see (1.4)) has the following counterpart for systems.

L^1-**lemma.** *Consider a C^2 solution u of (1.1), and write*

$$\partial_x u(x,t) = \sum_1^N w_j(x,t) r_j(u(x,t)).$$

Then the w_j satisfy a system

$$(2.5) \qquad L_i w_i = \Sigma \gamma_{ijk}(u) w_j w_k, \quad i = 1, \ldots, N,$$

where $L_i = \partial_t + \lambda_i(u(x,t))\partial_x$. If we define $\Gamma_{ijk}(u)$ by the equality

$$(2.6) \quad \sum_{j,k} \gamma_{ijk}(u) w_j w_k + w_i \Sigma w_k r_k(u) \cdot \partial_u \lambda_i(u) = \sum_{j,k} \Gamma_{ijk}(u) w_j \supset w_k,$$

the coefficients Γ_{ijk} have the following properties:

(i) *$\Gamma_{ijj} = 0$, and $\gamma_{ijj} = 0$ for $j \neq 0$.*

(ii) *If D_i is a domain bounded by an interval $[a,b]$ in $\{t = 0\}$, two segments of integral curves of L_i through $(a,0),(b,0)$ and an arc γ transverse to L_i, then*

$$(2.7) \quad \int_\gamma |w_i(dx - \lambda_i(u)dt)| \leq \int_a^b |w_i| dx + \int_{D_i} |\Sigma \Gamma_{ijk} w_j w_k| \, dx \, dt.$$

PROOF OF THE LEMMA.

a. It is obvious that

$$\partial_t z + A(u)\partial_x z = q_1(z), \quad z = \partial_x u,$$

where q_1 is a quadratic form in z with coefficients depending on u. Moreover,

$$\partial_x z = \Sigma r_i \partial_x w_i + q_2(w)$$

with q_2 a quadratic form in $w = (w_1, \ldots, w_N)$ (and coefficients depending on u) and similarly for $\partial_t z$; thus we obtain (2.5).

b. We now compute

$$d\big(w_i(dx - \lambda_i(u)dt)\big) = \Big\{\partial_t w_i + \lambda_i(u)\,\partial_x w_i + w_i\partial_x(\lambda_i(u))\Big\}dt\,dx$$

(2.8)
$$= \Big\{\Sigma\Gamma_{ijk}(u)\,w_j\,w_k\Big\}dt\,dx$$

by the definition of the Γ_{ijk}. If we take for u a j–simple wave, we have seen in (2.3) and (2.4) that $w_i = 0$ for $i \neq j$, $w_j = \partial_x\zeta$; hence $\Gamma_{ijj} = 0$ in this case. Since we can find a simple wave with $\partial_x\zeta \neq 0$ taking a given value at a given point, we get point (i) of the lemma.

c. Note that $|w_i|$ is Lipschitz continuous, hence differentiable almost everywhere with differential $\varepsilon d w_i$, ε being the sign of w_i. Hence we can write (2.8) in the form

$$d(|w_i|(dx - \lambda_i(u)dt)) = \varepsilon\Big\{\Sigma\Gamma_{ijk}\,w_j\,w_k\Big\}dt\,dx\,.$$

Using Stokes' formula for Lipschitz continuous forms, one obtains (2.7).

3. The case of 2×2 systems

3.1. The following theorem shows that blowup always occurs in this case for smooth compactly supported initial data, unless the eigenvalues are linearly degenerate.

Theorem 3.1. *Consider a 2×2 system in diagonal form*

$$\partial_t w + \Lambda(w)\partial_x w = 0$$

with initial data $w^0 \in C_0^\infty[a,b]$. Denote by $[m_i, M_i]$ the range of w_i^0 $(i = 1, 2)$, and set

$$\mu_1 = \max_{[m_2,M_2]}\lambda_1(0,w_2), \quad \mu_2 = \min_{[m_1,M_1]}\lambda_2(w_1,0).$$

We make on the eigenvalues and on the data the following assumptions:

(i) $\mu_1 < \mu_2$,

(ii) We have

$$\partial_{w_1}\lambda_1(\bar{w}_1,0) \neq 0$$

or

$$\partial_{w_2}\lambda_2(0,\bar{w}_2) \neq 0$$

for some $\bar{w}_i \in [m_i, M_i]$. Then the lifespan of w is finite.

PROOF OF THEOREM 3.1.

a. Assume $\bar{T} = \infty$. Then the values of w_i are always contained in the interval $[m_i, M_i]$. Let Γ^j_x the j−characteristic issued from $(x, 0)$. On Γ^1_b, we have $w_1 = 0$, hence $x \leq b + t\mu_1$. Similarly, on Γ^2_a, we have $w_2 = 0$ and $x \geq a + t\mu_2$. Thus, at time $T_0 = (b-a)(\mu_2 - \mu_1)^{-1}$, these two characteristics have already crossed each other, and the w_i are supported in disjoint strips for $t > T_0$.

b. Suppose that it is $\lambda_2(0, \cdot)$ which is not constant. The function $\lambda_2(0, w_2(x, T_0))$ is smooth and equal to $\lambda_2(0, 0)$ outside the interval image of $[a, b]$. If it were constant, this would imply that λ_2 is a constant as a function of w_2, because $w_2(x, T_0)$ takes on all values in the interval $[m_2, M_2]$. Thus there is an x for which $\partial_x\{\lambda_2(0, w_2(x, T_0))\} < 0$. According to the study of the scalar case in section 1, this implies finite time blowup for the scalar equation $\partial_t w_2 + \lambda_2(0, w_2)\partial_x w_2 = 0$, a contradiction.

3.2. To investigate further the exact nature of the blowup, we must distinguish two situations:

(i) The blowup occurs after the decoupling of w_1 and w_2 (that is, after Γ^1_b has crossed Γ^2_a): then we are left with a scalar situation where the mechanism is well understood ("geometric blowup").

(ii) The blowup occurs before the decoupling.

We will now show that, in general, the blowup is of a geometric nature also in this second situation.

Changing slightly the notations, consider a system in diagonal form

$$\partial_t w + \Lambda(w)\partial_x w = 0,$$

and assume given a smooth bounded solution w in a rectangle

$$\{(x, t), \ |x| < M, \ -T_0 \leq t < 0, \ M > 0\}.$$

We make the two following assumptions:

H_1 (Geometric assumption)

(i) There exist two 1-characteristics Γ_α^1 and Γ_β^1 issued from points

$$\alpha = (a, -T_0), \quad \beta = (b, -T_0), \quad a < b,$$

reaching points

$$\alpha' = (a', 0), \quad \beta' = (b', 0), \quad a' < 0 < b'$$

and such that the 2-characteristic Γ_α^2 from α intersects Γ_β^1 in α'', a point of negative ordinate.

(ii) There exists a 2-characteristic from a point γ of Γ_α^1 with negative ordinate, reaching $\gamma'' = (c'', 0)$, $c'' < 0$.

(iii) There exists a 1-characteristic from $\gamma' = (c, -T_0)$ reaching γ''.

H_2 (Analytic assumption)

In the polygon ω bounded by $\Gamma_\alpha^1, \Gamma_\beta^1, \{t = -T_0\}$ and $\{t = 0\}$, we have $|\partial_x w_2| \leq C$.

Recall from Chapter I that the blowup system (along λ_1) of a diagonal system is given by

$$(3.1) \qquad \partial_T \phi = \lambda_1(v), \; \partial_T v_1 = 0, \; (\lambda_2 - \lambda_1)(v) \partial_X v_2 + \partial_X \phi \partial_T v_2 = 0,$$

where v and ϕ are related to w by the formal change

$$(3.2) \qquad x = \phi(X, T), \; t = T, \; w(\phi(X, T), T) = v(X, T).$$

From the solution w in ω we obtain the solution (ϕ, v) of (3.1) in

$$\Omega = \{(X, T), a \leq X \leq b, -T_0 \leq T < 0\}$$

by defining $\phi(X, T)$ as the abscissa of the point of ordinate T on the 1-characteristic issued from $(X, -T_0)$.

We can now state our theorem.

Theorem 3.2. *Consider $w \in C^\infty$ a solution satisfying the assumptions H_1 and H_2. Let (ϕ, v) be the corresponding solution of (3.1) in Ω. Then (ϕ, v) can be extended as a smooth solution $(\tilde{\phi}, \tilde{v})$ of (3.1) in a domain $\tilde{\Omega}$ containing Ω and the segment*

$$\{(X, T), c < X < b, T = 0\}$$

in its interior.

Before giving the proof of Theorem 3.2, let us comment on assumption H_2 in a simplified example.

Consider Burgers' equation coupled with another equation

$$\partial_t w_1 + w_1 \partial_x w_1 = 0, \quad \partial_t w_2 + \lambda_2(w_1, w_2) \partial_x w_2 = 0$$

and assume that w_1 blows up at the origin.

Along a 2-characteristic, we have

(3.3)
$$(\partial_t + \lambda_2 \partial_x)(\partial_x w_2) + \partial_2 \lambda_2(w_1, w_2)(\partial_x w_2)^2 = -\partial_1 \lambda_2(w_1, w_2) \partial_x w_1 \partial_x w_2.$$

Since

$$(\partial_t + \lambda_2 \partial_x) w_1 = (\lambda_2 - \lambda_1) \partial_x w_1,$$

the integral $\int^0 |\partial_x w_1| ds$ along a 2-characteristic is finite and the presence of $\partial_x w_1$ in the right-hand side of (3.3) does not imply blowup of w_2.

We believe that, in general, $\partial_x w_1$ and $\partial_x w_2$ do not both blow up at time $\bar{T}$.

Note that this situation is in contrast with the similar situation of a semilinear system considered in Chapter IA2.

PROOF OF THEOREM 3.2.

1. Let us denote by $A = (X_A, T_A)$ the point of image α by the application

$$(X, T) \mapsto (\phi(X, T), T),$$

and so on. Since $\partial_T v_1 = 0$ and $v_1(X, -T_0) = w_1(X, -T_0)$, we consider the function v_1 as known.

Since

$$\partial_T v_2(X, T) = (\partial_t w_2 + \partial_T \phi \partial_x w_2)(\phi, T) = (\lambda_1 - \lambda_2) \partial_x w_2,$$

we deduce from H_2 that $|\partial_T v_2| \leq C$ in Ω. Because

$$\partial_T \partial_X \phi = \partial_1 \lambda_1 \partial_X v_1 - \frac{\partial_2 \lambda_1}{\lambda_2 - \lambda_1} \partial_X \phi \partial_T v_2,$$

we obtain $|\partial_X \phi| \leq C$ in Ω. Moreover, we obtain from (3.1), with $k = (\lambda_2 - \lambda_1)^{-1} \partial_X \phi$,

(3.4) $\partial_X v_2 + k \partial_T v_2 = 0, \quad (\lambda_2 - \lambda_1) \partial_T k + \partial_2 \lambda_2 k \partial_T v_2 - \partial_1 \lambda_1 \partial_X v_1 = 0.$

2. In the polygon $\Omega_1 = A\,A'\,B'\,A''\,A$ of the (X,T) plane, we consider the 2-characteristic from (a, T'): the ordinate of its point of abscissa X', which exists for X' in a maximal interval $[a, \zeta(T'))$ (for a nonincreasing function $\zeta(T') \leq b$), will be denoted by $\psi(X', T')$. Thanks to H_1, $\zeta(T') = b$ for $-T_0 \leq T' \leq -T_0 + \varepsilon_1$ and $\zeta(T_C) = c_1$ for some $c_1, a < c_1 \leq c$.

We thus define an application

$$(X', T') \mapsto (X, T), \quad X = X', \quad T = \psi(X', T')$$

which is a bijection from the domain

$$\bar{D}_1 = \left\{ ([a, b] \times [-T_0, -T_0 + \varepsilon_1]) \cup (a \leq X' \leq \zeta(T'), -T_0 + \varepsilon_1 \leq T' < 0) \right\}$$

onto Ω_1.

3. We define now in $\bar{D}_1$ the functions ℓ and h by

$$k(X', \psi(X', T')) = \ell(X', T'), \quad v_2(X', \psi(X', T')) = h(X', T')$$

and ℓ, h, ψ satisfy the system

(3.5) $$\ell = \partial_{X'} \psi, \quad \partial_{X'} h = 0,$$

(3.6)
$$\partial_{T'} \ell + \frac{\partial_2 \lambda_2}{\lambda_2 - \lambda_1} (v_1(X'), h) \ell \partial_{T'} h - \frac{\partial_1 \lambda_1}{\lambda_2 - \lambda_1} (v_1(X'), h) \partial_{T'} \psi \partial_X v_1(X') = 0.$$

The boundary conditions are the following:

(i) On $X' = a$, $h = v_2$, $\psi(a, T') = T'$.

(ii) On $T = -T_0$, ψ is known from its definition, with $\psi(a, -T_0) = -T_0$.

The key point is this: if, in (3.6), we consider v_1 and h as known and replace ℓ by $\partial_{X'} \psi$, we obtain a **linear** equation on ψ. Taking into account the above boundary conditions, we see that we have to deal with a **linear Goursat problem**.

4. Since the values of ψ are compatibles and smooth on the closed segments $\{X' = a, -T_0 \leq T \leq T_C\}$ and $\{T' = -T_0, a \leq X' \leq b\}$, there exists a smooth solution $\tilde{\psi}$ extending ψ in the closed rectangle

$$\bar{R} = \left\{ (X', T'), a \leq X' \leq b, -T_0 \leq T \leq T_C \right\}.$$

5. The domain

$$\bar{D}_2 = \bar{D}_1 \cap \{T' \leq T_C\}$$

is contained in $\bar{R}$. We claim that

$$\partial_{T'} \psi(X', T') \neq 0, \quad (X', T') \in \bar{D}_2.$$

Actually, in the interior of $\bar{D}_2$, we have

$$\partial_T v_2(X', \psi) \partial_{T'} \psi = \partial_{T'} h.$$

Since $\partial_T v_2$ is bounded by (1), if $\partial_{T'} \tilde{\psi} = 0$ at some point (X_1', T_1'), we have $\partial_{T'} h(X', T_1') = 0$ for all X'. Hence (3.6) becomes an homogeneous ODE on $\partial_{T'} \tilde{\psi}$ along the line $T' = T_1'$, and our assumption leads to $\partial_{T'} \psi(a, T_1') = 0$, which is impossible.

6. We deduce from (5) that there are extensions $\tilde{v}_2, \tilde{k}$ of v_2, k (satisfying (3.1)) to a new domain containing the polygon $ACC''B'BA$ and the path $CC''B'B$ in its interior.

We choose now $\tilde{\phi}$ satisfying

$$\partial_X \tilde{\phi} = (\lambda_2 - \lambda_1)\tilde{k}, \quad \tilde{\phi}(b, T) = \phi_1(T).$$

Here $\phi_1(T)$ is an extension of $\phi(b, T)$ to an interval $[-T_0, \varepsilon]$ for which $\partial_T \phi_1 = \lambda_1$.

This construction defines $\tilde{\phi}$ on the rectangle

$$[c_1, b] \times [-T_0, \varepsilon]$$

for ε small enough, which completes the proof. $\qquad\qquad \diamondsuit$

We can roughly summarize the proof by saying that the problem is turned into a linear problem by taking the unknown functions w as new independant variables; this is the so called **"hodograph method"** (see for instance [CFr]).

4. General systems with small data

We consider here a $N \times N$ system for which we have performed the normalization explained in (2), that is, the j-axis in $\mathbb{R}_u^N$ is the integral curve of $r_j(u)$ through the origin.

Consider a solution u of such a system with initial value of the form

$$(4.1) \qquad u(x, 0) = u^0(x, \varepsilon) = \varepsilon u_1^0 + \varepsilon^2 u_2^0 + \cdots$$

where $u^0(x, \varepsilon)$ is a smooth function for $x \in \mathbb{R}$, $0 \leq \varepsilon \leq \varepsilon_0$, supported in $a \leq x \leq b$ (a, b independent of ε).

Let Γ_x^j be the integral curve $(x_j(x,t),t)$ of $L_j = \partial_t + \lambda_j(u)\partial_x$ from $(x,0)$, defined for $t < \bar{T}_\varepsilon$ (the lifespan of u); R_j is the strip limited by $\Gamma_a^j, \Gamma_b^j, \{t = 0\}$.

Setting

$$\partial_x u(x,t) = \Sigma w_j(x,t) r_j(u(x,t))$$

as in the L^1-lemma, we introduce the quantities

$$(4.2) \qquad\qquad I(t) = \max_{0 \le s \le t} \sum_j \int |w_j(x,s)| dx$$

and

$$(4.3) \qquad\qquad S(t) = \max_{j, 0 \le s \le t} \max_{(x,s) \notin R_j} |w_j(x,s)|.$$

We expect $I(t)$ to be $O(\varepsilon)$ and $S(t)$ to be $O(\varepsilon^2)$, because the strip R_j is the "natural domain" for w_j to live, at least at first approximation.

We first prove the following striking result.

Theorem 4.1. *Let $M > 0$ be some fixed number and assume that the solution u is smooth for $0 \le t < T(\varepsilon)$, with $\varepsilon T(\varepsilon) \le M$. Then there are constants $I_0, S_0, \delta_0, \varepsilon_1 > 0$ such that, for $0 \le t < T(\varepsilon)$ and $0 < \varepsilon \le \varepsilon_1$*

$$(4.4) \qquad\qquad I(t) \le I_0 \varepsilon, \quad S(t) \le S_0 \varepsilon^2, \quad |u| \le \delta_0.$$

PROOF OF THEOREM 4.1.

1. We will argue as follows (a procedure called "induction on time"). Suppose that we have found constants I_0, S_0, δ_0 such that

(i) We have, for some $T' \le T(\varepsilon)$ and ε small enough,

$$(4.5) \qquad\qquad I(t) < I_0 \varepsilon, \quad S(t) < S_0 \varepsilon^2, \quad |u| < \delta_0$$

for $t < T'$.

(ii) We can prove that whenever (4.5) holds for some $T'' < T(\varepsilon)$ and $t < T''$, it also holds for $t = T''$.

Then (4.5) is true for $0 \le t < T(\varepsilon)$ (because it can never stop being true).

2. Fix any T' (independent of ε). We first prove that (i) holds for T' (for δ_0 arbitrary and appropriate I_0, S_0).

Elementary results show that, for $t \leq T'$,

$$|\partial_{x,t}^\alpha u| \leq C_\alpha \varepsilon.$$

Since we have the identity

$$L_i w_i = \Sigma \gamma_{ijk} w_j w_k,$$

we also have

$$w_i(x,t) = O(\varepsilon^2), \quad (x,t) \notin R_i.$$

Thus (i) follows for ε small enough and appropriate I_0, S_0.

3. Choose moreover I_0 such that $I(0) \leq \frac{I_0}{2}\varepsilon$; δ_0 such that $|\lambda_j - \lambda_i| \geq c > 0$ for $|u| \leq \delta_0$. Then, for $t > T_0 = \frac{(b-a)}{c}$, the strips R_i are disjoints. In the following estimates, we always distinguish what happens before T_0 from what happens after T_0.

We use now the L^1-lemma to estimate I: since there are no squares in the quadratic terms ($\Gamma_{ijj} = 0$), **for** $s > T_0$, we can estimate each integral

$$\int |w_j(x,s)||w_k(x,s)|dx$$

by $I(s) S(s)$. In fact, the strips R_j and R_k being disjoint, we take w_j out of the integral if $x \notin R_j$, and w_k out of the integral if $x \in R_j$.

Thus we obtain for $s \leq t < T''$

$$\Sigma \int |w_i(x,s)| dx \leq \frac{I_0}{2}\varepsilon + C\varepsilon^2 + CtI(t)S(t) \leq$$

$$\leq I_0\varepsilon\left[\frac{1}{2} + \left(\frac{C}{I_0} + CMS_0\right)\varepsilon\right].$$

Taking ε small enough, we get $I(t) \leq \frac{2I_0}{3}\varepsilon$.

4. Since u has compact support for fixed t, the result of (3) implies $|u| \leq C\varepsilon$; hence in particular $|u| \leq \frac{\delta_0}{2}$ for ε small enough.

5. To evaluate $w_i(x,s)$, $(x,s) \notin R_i$, $s \leq t < T''$, we draw back the integral curve of L_i through (x,s) and use (2.5). Remark that this curve never meets R_i. The initial value of w_i is 0 and the portion of integral for $0 \leq t \leq T_0$ is bounded by $C\varepsilon^2$.

Consider the integral from T_0 to s of a term $\gamma_{ijk} w_j w_k$:

(i) For the portion of curve not meeting R_j or R_k, the integral is bounded by $C\frac{M}{\varepsilon}S(s)^2 \le CM S_0^2 \varepsilon^3$.

(ii) The portion ℓ_{ij} of curve contained in R_j ($j \ne i$) is of finite length; since $k \ne j$ (because $\gamma_{ijj} = 0$ for $i \ne j$ according to the L^1-lemma) the corresponding integral can be estimated by

$$S(s) \int_{\ell_{ij}} |w_j|.$$

To estimate this last piece of integral, we use again (2.7); we get as in (3),

$$C \int_{\ell_{ij}} |w_j| \le \frac{I_0}{2}\varepsilon + C\varepsilon^2 + CM S_0 I_0 \varepsilon^2.$$

Collecting the terms, we obtain finally

$$|w_i(x,s)| \le C\varepsilon^2 + CM S_0^2 \varepsilon^3 + S_0 \varepsilon^3 (I_0 + C\varepsilon + CM S_0 I_0 \varepsilon).$$

If $S_0 \ge 2C$, we obtain $S(t) \le \frac{2S_0}{3}\varepsilon^2$ for ε small enough, which completes the proof. $\Diamond$

Theorem 4.1 makes it possible to reduce lower and upper estimates of the lifespan $\bar{T}_\varepsilon$ to corresponding estimates for solutions of ODE. The idea comes from (2.5): along an integral curve of L_i in R_i, the main term of the sum in the right-hand side of (2.5) is $\gamma_{iii} w_i^2$.

We first state the following lemma.

ODE Blowup lemma. *Let w be a solution in $[0, T]$ of the ODE*

$$(4.6) \qquad \frac{dw}{dt} = a_0(t)w^2 + a_1(t)w + a_2(t)$$

with a_j continuous and $a_0 \ge 0$. Let

$$(4.7) \qquad K = \left(\int_0^T |a_2(t)|\, dt \right) \exp \int_0^T |a_1(t)|\, dt.$$

If $w(0) > K$, we have

$$(4.8) \qquad \int_0^T a_0(t)\, dt < (w(0) - K)^{-1} \exp \int_0^T |a_1(t)|\, dt.$$

PROOF OF THE LEMMA.

a. Setting

$$w(t) = W(t) \exp \int_0^t a_1(s) ds,$$

we obtain the equation

$$\frac{dW}{dt} = A_0(t) W^2 + A_2(t),$$

where

$$A_0(t) = a_0(t) \exp \int_0^t a_1(s) ds, \quad A_2(t) = a_2(t) \exp - \int_0^t a_1(s) ds.$$

Thus, if the lemma is true for $a_1 \equiv 0$, it is true in general since

$$W(0) = w(0) > K \geq \int_0^T |A_2(t)| dt$$

and

$$\left(\int_0^T a_0(t) dt \right) \exp - \int_0^T |a_1(t)| dt \leq \int_0^T A_0(t) dt.$$

b. The solution w_1 of the model equation

$$\frac{dw_1}{dt} = a_0(t)(w_1 - K)^2, \quad w_1(0) = w(0)$$

can be computed explicitly and satisfies

$$(w_1(t) - K)^{-1} - (w(0) - K)^{-1} = - \int_0^t a_0(s) ds.$$

Thus if w_1 exists in $[0, T]$,

(4.9) $$\int_0^T a_0(t) dt < (w(0) - K)^{-1}.$$

c. With $w_2(t) = \int_0^t |a_2(s)| ds$ we have now

$$\frac{d(w_1 - w_2)}{dt} = a_0(t)(w_1 - K)^2 - |a_2(t)| \leq a_0(t)(w_1 - w_2)^2 + a_2(t)$$

and $(w_1 - w_2)(0) = w(0)$. It follows that as long as w_1 exists, $w_1 - w_2 \leq w$. Thus w_1 cannot blow up for $t \leq T$, hence exists on $[0, T]$ and (4.9) gives then (4.8). ◇

We can now prove the following theorem.

Upper Bound Theorem 4.2. *Assume that the $j-$eigenvalue λ_j is genuinely nonlinear at 0, that is (with the already performed normalization)*

$$\partial_j \lambda_j(0) \neq 0.$$

Then

$$(4.10) \qquad \limsup \varepsilon \bar{T}_\varepsilon \leq \{ \max_x -(u_1^0)'_j(x)\partial_j\lambda_j(0)\}^{-1} = M_j.$$

PROOF OF THEOREM 4.2.

   a. Let B be strictly bigger than the right-hand side M_j of (4.10), and assume $\varepsilon \bar{T}_\varepsilon > B$. We can then use Theorem 4.1 with $\varepsilon T(\varepsilon) = B$.
   Let x_0 be such that

$$-(u_1^0)'_j(x)\,\partial_j\lambda_j(0)$$

is maximum at x_0 and consider (2.5) as an ODE on $z_j = \eta w_j$ ($\eta = \text{sgn}\,(-\partial_j\lambda_j(0))$) on $\Gamma_{x_0}^j$.
   With the notations of the ODE Blowup Lemma, we have now

$$a_0 = \eta\gamma_{jjj}(u), \quad a_1 = 2\sum_{i \neq j} \gamma_{jji} w_i, \quad a_2 = \eta \sum_{i,k \neq j} \gamma_{jik}\, w_i\, w_k.$$

   b. From (4.4) we deduce that in fact

$$|u| \leq C\varepsilon.$$

Hence

$$w_j(x_0, 0) = (u_1^0)'_j(x_0)\varepsilon + O(\varepsilon^2)$$

and

$$z_j(x_0, 0) = \varepsilon(M_j|\partial_j\lambda_j(0)|)^{-1} + O(\varepsilon^2) > 0.$$

On the other hand, by (2.6) and $\Gamma_{jjj} = 0$, we have

$$\gamma_{jjj}(0) = -\partial_j\lambda_j(0),$$

hence $a_0(u) = |\partial_j\lambda_j(0)| + O(\varepsilon)$.
   Finally, by Theorem 4.1,

$$a_1 = O(\varepsilon^2), \quad a_2 = O(\varepsilon^4).$$

c. We apply the lemma on the interval $[0, B\varepsilon^{-1}]$; from (b) we have $K = O(\varepsilon^3)$; thus the lemma gives

$$\varepsilon(1 + O(\varepsilon)) \int_0^{B\varepsilon^{-1}} a_0(t)dt < M_j|\partial_j\lambda_j(0)| + O(\varepsilon),$$

yielding a contradiction for ε small enough.   ◇

Clearly, if all the eigenvalues of the system are genuinely nonlinear, Theorem 4.2 yields the upper bound

$$(4.11) \qquad\qquad \limsup \varepsilon\bar{T}_\varepsilon \leq \inf M_j.$$

Up to now, we have concentrated our attention on upper bounds for the lifespan, since our main interest is in blowup.

By using, instead of the ODE Blowup Lemma, a similar ODE Existence Lemma, one can prove that the upper bound obtained in the proof of Theorem 4.2 for the lifespan of w_j is asymptotically sharp. Once we have gained control of $|\partial_x u|$, the blowup criterion from Chapter II gives a lower bound of the lifespan.

Summing up, we have the following statement.

Theorem 4.3. *If all the eigenvalues of the system are genuinely nonlinear, the lifespan $\bar{T}_\varepsilon$ of the solution u is asymptotically given by*

$$(4.12) \qquad\qquad \lim \varepsilon\bar{T}_\varepsilon = \inf_j M_j.$$

We finish this section with two remarks:

(i) It is a consequence of Theorem 4.1 and 4.2 that the solution u itself remains bounded for $t < \bar{T}_\varepsilon$ (in fact, $|u| \leq C\varepsilon$). For general systems of the form (1.1) and smooth compactly supported data (not necessarily small), it does not seem to be known whether the solution u remains bounded and only ∇u becomes infinite. However, this seems to be a general belief.

(ii) In the situation of Theorem 4.3, if the M_j are distinct (a "generic" case), only $|w_j|$ (j such that $M_j = \inf M_k$) becomes infinite, at time $\bar{T}_\varepsilon$ and in the strip R_j. This is very similar to the way a blowup solution along λ_j blows up (compare in particular Theorem 5.1 of Chapter I with the estimates (4.4)). We will see in the next chapter that the solution is actually a "blowup solution".

5. Rotationally invariant wave equations

We consider now the equation

$(5.1)_a$ $$\partial_t^2 u - c^2(\partial_t u)\Delta u = 0, \quad x \in \mathbb{R}^3,$$

where

(5.2) $$c(0) = 1, \quad c'(0) > 0$$

and

$(5.1)_b$ $$u(x,0) = \varepsilon f(x), \quad \partial_t u(x,0) = \varepsilon g(x).$$

We assume that f and g are smooth **radial** functions (that is, smooth functions of $|x|^2$), supported in $|x| \leq M$. Because of the structure of (5.1), smooth solutions with radial data are also radial, that is, functions of t and $|x|^2$.

The Laplace operator in polar coordinates $\left(r = |x|, \, \omega = \frac{x}{r}\right)$ is given by

$$\Delta = \partial_r^2 + \frac{n-1}{r}\partial_r - r^{-2}\Delta_\omega.$$

It is thus convenient to introduce the function $v = ru$ (viewed as an odd function of r) which satisfies

(5.3) $$\partial_t^2 v - c^2\left(\frac{1}{r}\partial_t v\right)\partial_r^2 v = 0$$

and, with an obvious abuse of notation,

(5.4) $$v(r,0) = \varepsilon r f(r), \quad \partial_t v(r,0) = \varepsilon r g(r).$$

Equation (5.3) can be reduced to a first order system by introducing $\partial_r v$ and $\partial_t v$ as new unknowns. We will then follow the proof of Theorem 4.2, step for step (L^1-lemma, Theorem 4.1 and ODE Blowup Lemma), to obtain the following theorem.

Theorem 5.1. *Let $\bar{T}_\varepsilon$ be the lifespan of the solution u of (5.1) with radial data $\varepsilon f, \varepsilon g$. Then*

(5.5) $$\limsup \varepsilon \ln \bar{T}_\varepsilon \leq \{\max_\rho c'(0)F''(\rho)\}^{-1},$$

where the "free profile" F is defined by

(5.6) $$F(\rho) = \frac{1}{2}\left(\rho f(\rho) - \int_{-\infty}^{\rho} \sigma g(\sigma)\,d\sigma\right).$$

Note that, because of the parity assumptions on f and g, F is of compact support and does not vanish identically unless f and g do.

The proof of Theorem 5.1 is divided into three steps.

Step 1. The L^1-lemma

We set

$$\partial^2_{rt} v = c(w_2 - w_1), \quad \partial^2_{rr} v = w_1 + w_2,$$

so that we obtain by a straightforward computation

$$(5.7) \qquad L_1 w_1 = \frac{cc'}{r} w_1(w_1 - w_2) + \frac{c'\partial_t u}{2r}(3w_1 + w_2),$$

$$(5.8) \qquad L_2 w_2 = \frac{cc'}{r} w_2(w_2 - w_1) - \frac{c'\partial_t u}{2r}(3w_2 + w_1).$$

These formula are the analogues of (2.5). Here of course,

$$L_1 = \partial_t + c\partial_r, \quad L_2 = \partial_t - c\partial_r.$$

To obtain the analogue of (2.8), we compute $d(w_i(dr \pm cdt))$, and find

$$(5.9) \qquad d(w_1(dr - cdt)) = \frac{c'\partial_t u}{2r}(w_1 + w_2)dt\,dr,$$

$$d(w_2(dr + cdt)) = -\frac{c'\partial_t u}{2r}(w_1 + w_2)dt\,dr.$$

Step 2. The estimates up to the blowup time

Define as in 4. the strip R as being limited by Γ^1_{-M}, Γ^1_M, $\{r = 0\}$ and $\{t = 0\}$, and introduce the quantities

$$I(t) = \max_{0 \le s \le t} \int_{(r,s) \in R} |w_1(r,s)|dr,$$

$$S(t) = \max_{2M \le s \le t} s^2 \max_{(r,s) \in R} |w_2(r,s)|,$$

$$U(t) = \max_{2M \le s \le t} s \max_{(r,s) \in R} |\partial_t u(r,s)|.$$

Remark here that we are dealing with a very large existence time, a fact which forces us to take the decay of the various derivatives of u into account in the definitions of I, S, U.

We will prove the following analogue of Theorem 4.1.

Theorem. *Let $C_0 > 0$ be some fixed number and assume that the solution u is smooth for $0 \le t < T(\varepsilon)$, with $\varepsilon \ln T(\varepsilon) \le C_0$. Then there are constants I_0, S_0, U_0, $\varepsilon_1 > 0$ such that, for $0 \le t < T(\varepsilon)$ and $0 < \varepsilon \le \varepsilon_1$*

(5.10) $$I(t) \le I_0 \varepsilon, \quad S(t) \le S_0 \varepsilon^2, \quad U(t) \le U_0 \varepsilon.$$

PROOF OF THE THEOREM.

a. We proceed by induction on time.

We have first to make sure that the characteristics in the strip R look for large time the same as they do for small time, that is

(i) For $(r, t) \in \Gamma_\lambda^1$, $|r - t - \lambda| \le C$.

(ii) For (r, t), $(r', t') \in \Gamma_\mu^2 \cap R$, $|r + t - \mu| \le C$ and $|t - t'| \le C$ (with constants independent of μ).

By induction, (5.10) implies

$$|c - 1| = |c(\partial_t u) - 1| \le \frac{\varepsilon C U_0}{t}$$

which gives

$$|r - t - \lambda| = \left| \int_0^t (c - 1) ds - \lambda \right| \le C\varepsilon + CU_0 \varepsilon \ln t + |\lambda| \le C + CU_0 C_0.$$

The first part of (ii) is similar, and

$$2|t - t'| \le |(t + r - \mu) - (t' + r' - \mu)| + |t - r| + |t' - r'|$$

gives the second statement.

b. For any fixed $T' > 2M$, there are constants I_0, S_0, U_0 such that (5.10) is true for small ε, because of (5.8). We assume moreover $I(2M) \le \frac{\varepsilon I_0}{2}$.

c. To estimate $I(t)$, we use the L^1-lemma in R between $\{s = 2M\}$ and $\{s = t\}$. By (5.9), we get

$$\int_{(r,t) \in R} |w_1(r, t)| \, dr \le \frac{\varepsilon I_0}{2} + \int_{2M \le s \le t, (r,s) \in R} \left| \frac{c' \partial_t u}{2r} (w_1 + w_2) \right| dr \, ds.$$

The second term is bounded by

$$\varepsilon C U_0 I(t) \int_{2M}^t \frac{ds}{s^2} + \varepsilon C U_0 S(t) \int_{2M \le s \le t, (r,s) \in R} \frac{dr \, ds}{r s^3},$$

hence by $\frac{\varepsilon I_0}{4}$ for ε small enough.

Finally, $I(t) \leq \frac{2\varepsilon I_0}{3}$ for ε small enough.

d. Noting that

$$(5.11) \qquad L_2(\partial_t u) = \frac{c}{r}(2cw_1 + \partial_t u),$$

we can estimate $\partial_t u(r, t)$ by integrating along the arc of Γ_μ^2 issued from (r, t) between (r, t) and the point (r', t') where Γ_μ^2 leaves R (where the solution and its derivatives are zero).

First we simplify (5.11) by introducing $z = \partial_t u \exp - \int^t \frac{c}{r} ds$, remarking that

$$\int^t \frac{c}{r} dt \leq C(1 + \varepsilon U_0).$$

e. We need now to estimate the integral of $|w_1|$ on an arc of Γ_μ^2 in R. Using the L^1-lemma and the fact that, along such an arc, $dr - c\,dt = -2c\,dt$, a computation analogous to (c) gives

$$\int_{\Gamma_\mu^2 \cap R} |w_1| \, dt \leq \frac{\varepsilon I_0}{2} + C U_0 I_0 \varepsilon^2 + C U_0 S_0 \varepsilon^3.$$

From (d) we obtain finally

$$t|z| \leq C\varepsilon I_0 + C U_0 I_0 \varepsilon^2 + C U_0 S_0 \varepsilon^3.$$

If we choose $U_0 = \nu I_0$, ν big enough and then ε small enough, we get finally $U(t) \leq \frac{2\varepsilon U_0}{3}$.

f. To estimate w_2, we proceed as before with $\partial_t u$, using (5.8) instead of (5.11).

We write (5.8) in the form $L_2 w_2 = aw_2 + b$, with

$$a = \frac{c'(2c(w_2 - w_1) - 3\partial_t u)}{2r}, \qquad b = -\frac{c'}{2r} w_1 \partial_t u.$$

We see that the integral of $|aw_2|$ is $O(\frac{\varepsilon^3}{t^2})$, while

$$\int |b| \, ds \leq \frac{\varepsilon C U_0}{t^2} \int |w_1| \, ds.$$

Taking (d) into account, we get $S(t) \leq \frac{2\varepsilon^2 S_0}{3}$ if we choose $\frac{S_0}{U_0 I_0}$ big enough and then ε small enough. $\diamond$

Step 3. ODE Blowup

We view now (5.7) as an ODE along Γ_λ^1 between $\{t = 2M\}$ and $\{t = T\}$. With the notations of the ODE Blowup Lemma, we have

$$a_2(t) = O\Big(\frac{\varepsilon^3}{t^4}\Big), \quad a_1(t) = O\Big(\frac{\varepsilon}{t^2}\Big),$$

hence the number K is $O(\varepsilon^3)$. Moreover, $a_0(t) = c'(0)\frac{1+O(\varepsilon)}{t+O(1)}$.

For $0 \le t \le 2M$, the solution v of (5.3), (5.4) can be approximated by the solution $\bar{v}$ of the equation $(\partial_t^2 - \partial_r^2)\bar{v} = 0$, with $v - \bar{v} = O(\varepsilon^2)$. Since, by an elementary computation,

$$\bar{v}(r,t) = \varepsilon F(r - t), \quad t > M,$$

we get

$$w_1(r, 2M) = \frac{1}{2}(\partial_{rr}^2 v - c^{-1}\partial_{rt}^2 v)(r, 2M) = \varepsilon F''(r - 2M) + O(\varepsilon^2).$$

If we choose the characteristic Γ_λ^1 such that it meets the line $\{t = 2M\}$ at $(r_0, 2M)$ with

$$F''(r_0 - 2M) = \max F''(\rho),$$

we obtain from the lemma

$$\varepsilon \, \ell n \, \bar{T}_\varepsilon \le (c'(0) \max F''(\rho))^{-1}(1 + O(\varepsilon)),$$

which finishes the proof of Theorem (5.1).   ◇

Here again, as in Section 4, we have concentrated on the proof of an upper bound for the lifespan. We will show in the next chapter that we have in fact **equality** in (5.6).

To conclude, let us emphasize once more that the essential ingredient of the proofs of Theorems 4.2 and 5.1 is the L^1-lemma; for space dimension $n \ge 2$, there is no analogue of the L^1 estimates for ∇u, and the situation is still poorly understood.

Notes

Most of the material in this chapter is due to F. John [Jo2], [Jo4], [Jo6], though the presentation here follows closely L. Hörmander [Hö2].

The generalities of Section 1, 2 and Section 4 come from [Jo2] and [Hö2], while Section 5 is taken from [Jo4] and [Hö2].

The survey [Jo6] covers the content of Sections 1, 2, 4 and 5.

In Section 3, Theorem 3.1 follows the presentation by A. Majda [Ma] of a theorem of S. Klainerman and himself, while Theorem 3.2 is due to the author [Al9].

CHAPTER V

Nonlinear Geometrical Optics
and Applications

Introduction

In the theory of linear PDE, the expression "geometrical optics" refers to the search of special high frequency solutions of a given PDE in the form

$$u(x, \nu) = a(x, \nu) \exp \nu \phi(x),$$

where ν is the frequency ($\nu \to \infty$), a the amplitude (a formal power series in $\frac{1}{\nu}$) and ϕ the phase.

Similar constructions have been used in many nonlinear problems; their common feature is that the solution has a certain given form called "ansatz", containing undetermined coefficients or functions and a small parameter.

We restrict ourselves here to the study of the Cauchy problem with small initial data of compact support. The term "nonlinear geometrical optics" refers then to an approximation technique which seems to have its origin in perturbation theory for ordinary differential equations (see Arnold [Ar] for instance). For data of size ε, this technique yields approximations valid on time intervals typically of size ε^{-1}, involving the "slow time" $\tau = \varepsilon t$ as an auxiliary variable.

In the framework of one dimensional systems studied in Chapter IV, we obtain in Section 1 a good description of the solution for $0 \le t \le A\varepsilon^{-1}$, for any A and ε satisfying

$$A < \lim \varepsilon \bar{T}_\varepsilon, \ 0 < \varepsilon \le \varepsilon_A.$$

On the one hand, this description is more detailed than what can be deduced from the estimates of Theorem 4.1 of Chapter IV; on the other hand, it is not uniformly valid up to the blowup time (in sharp contrast with Theorem 4.1), and in particular it does not allow **by itself** to prove actual blowup of the solution.

The technique of nonlinear geometrical optics can be used also for several space variables; despite its above mentioned drawbacks, it seems to be the only (presently known) way of getting insight into what is actually happening, thus guiding the search for relevant estimates.

The examples handled in the literature are the compressible Euler equations (only in the 2-dimensional axisymetric case, see Alinhac [Al5]) and quasilinear wave equations (see John [Jo5], Hörmander [Hö1], [Hö2], Alinhac [Al1], [Al2], [Al3]).

In this latter case, discussed in Section 2, the nonlinear geometrical optics expansions suggest the blowup behavior of the solution and make transparent the role played by the "null conditions".

Finally, we present in Section 3 some further results on the wave equation in dimension two, showing the existence of an "asymptotic life-span".

1. Quasilinear systems in one space dimension

We discuss again the case already considered in Chapter IV

$$(1.1) \quad \partial_t u + A(u)\partial_x u = 0, \quad u(x,0) = u_0(x) = \varepsilon\, u_0^{(1)}(x) + \varepsilon^2\, u_0^{(2)}(x) + \dots,$$

where the slight change of notations is due to the fact that we will have to consider separately the components v_j of vectors $v = (v_1, \dots, v_N) \in \mathbb{R}^N$.

We take $u_0 \in C^\infty$ supported in $[a, b]$.

Here, we again assume the normalization of coordinates in the u-space achieving that the j−axis in an integral curve of $r_j(u)$ (see Chapter IV 2). It means, with e_j the j−basis vector and $A = (a_{jk})$, that

$$a_{jk}(te_k) = \delta_{jk}\lambda_k(te_k),$$

which implies in particular

$$(1.2) \quad \partial_k^q a_{jk}(0) = 0, \quad j \neq k, \quad q \in \mathbb{N}.$$

This will be essential to obtain the simple statements of Theorem 1.1 below.

1.1. Formal analysis

To solve (1.1), we expand $A(u) = \Sigma_\alpha A_\alpha u^\alpha$ and look formally for

$$u(x, t, \varepsilon) = u(x, t) = \sum_{p \geq 1} \varepsilon^p u^{(p)}(x, t).$$

With $L = \partial_t + A(0)\partial_x$, we choose the $u^{(p)}$ satisfying the sequence of equations

$(1.3)_1$ $\qquad\qquad Lu^{(1)} = 0, \qquad u^{(1)}(x, 0) = u_0^{(1)}(x),$

$(1.3)_p$ $\qquad\qquad Lu^{(p)} = f^{(p)}, \quad u^{(p)}(x, 0) = u_0^{(p)}(x),$

where $f^{(p)}$ is a polynomial expression of the $u_j^{(q)}$ ($q \leq p-1$), which can be easily computed from the expansion of A.

To describe the supports and the structure of the terms $u_j^{(p)}$, we introduce the following notations:

$$c = \inf_{1 \leq j \leq N-1}(\lambda_{j+1}(0) - \lambda_j(0)), \quad \gamma = \frac{(\lambda_{N-1}(0) - \lambda_2(0))}{c},$$

$$\alpha_j = (\lambda_j(0), 1),$$

$$K_T = \Big\{(x, t), a + \lambda_2(0)t \leq x \leq b + \lambda_{N-1}(0)t, 0 \leq t \leq T\Big\},$$

(1.4) $\qquad\qquad T_1 = 0, \ T_p = \frac{(b-a)}{c} \sum_{0 \leq q \leq p-2}(1+\gamma)^q, \ p \geq 2,$

$$K_p = K_{T_p},$$

$$K_p^j = (K_p + \mathbb{R}\alpha_j) \cap \{t \leq T_p\}.$$

The following theorem gives the form of the terms $u^{(p)}$.

Theorem 1.1. *For all $p \geq 1$ and $1 \leq j \leq N$ we have*
(1.5)
$$u_j^{(p)}(x, t) = \sum_{0 \leq q \leq p-1} t^q v_{jq}^{(p)}\ (\sigma_j(x, t)) + r_j^{(p)}(x, t), \sigma_j(x, t) = x - \lambda_j(0)t,$$

where $r_j^{(p)}$ is a C^∞ function supported in K_p^j and $u_j^{(p)}$ is supported in $K_p + \mathbb{R}_+\alpha_j$.

PROOF OF THEOREM 1.1.

a. We have immediately

$$u_j^{(1)} = v_{j0}^{(1)}(\sigma_j), \quad v_{j0}^{(1)} = (u_0^{(1)})_j, \quad r_j^{(1)} \equiv 0,$$

which is (1.5) for $p = 1$.

b. Assume the theorem already proved for $q \leq p - 1$ and consider now a monomial in $f_j^{(p)}$; there are two cases:

(i) This monomial contains only terms $u_k^{(q)}$ and their derivatives for **the same** k.

(ii) This monomial contains at least two terms $u_k^{(q)}$, $u_{k'}^{(q')}$, $k \neq k'$.

c. We analyze case (i) first (called "resonant interaction").

The monomial must contain a coefficient $\partial_k^s a_{jk}(0), s \neq 0$. But this is zero, by (1.2), if $k \neq j$.

d. In case (ii) (called "non resonant"), the support of the monomial is contained in

$$(K_q + \mathbb{R}_+ \alpha_k) \cap (K_{q'} + \mathbb{R}_+ \alpha_{k'}) \subset (K_{p-1} + \mathbb{R}_+ \alpha_k) \cap (K_{p-1} + \mathbb{R}_+ \alpha_{k'}) = S_{kk'}.$$

We show that

(1.6) $$S_{kk'} \subset K_p.$$

Let us denote by I_p the top segment of K_p

$$I_p = \left\{ (x,t), t = T_p, a + \lambda_2(0)T_p \leq x \leq b + \lambda_{N-1}(0)T_p \right\}$$

and by ℓ_p its length.

We see easily that (1.6) is true for $p = 2$. Now we also see that $S_{kk'}$ is contained in $(I_{p-1} + \mathbb{R}_+ \alpha_k) \cap (I_{p-1} + \mathbb{R}_+ \alpha_{k'})$; by the same reasoning as for $p = 2$, we get

$$S_{kk'} \subset K_T,$$

where $T = T_{p-1} + \frac{\ell_{p-1}}{c}$. Noting that $\ell_{p-1} = (b - a) + c\gamma T_{p-1}$, we obtain $T = T_p$.

e. Using the recurrence hypothesis and (c), (d), we see that $f_j^{(p)}$ is a sum of terms either

(i) with compact support in K_p (for nonresonant interactions),

(ii) or with support in $K_{p-1} + \mathbb{R}_+ \alpha_j$ and of the form

$$t^q w(\sigma_j) + r,$$

r being itself supported in K_p^j. To determine q, we observe that if the term is obtained as a product of ℓ terms of the same type corresponding to $p = p_{i_1}, \ldots, p_{i_\ell}$, we have

$$p_{i_1} + \ldots + p_{i_\ell} = p, \quad q_{i_j} \le p_{i_j} - 1,$$

hence, because $\ell \ge 2$,

$$q = q_{i_1} + \ldots + q_{i_\ell} \le p - \ell \le p - 2.$$

f. When we integrate from 0 to t along $L_j = \partial_t + \lambda_j(0)\partial_x$ a term supported in $K_p + \mathbb{R}_+ \alpha_j$, we obtain a term supported in the same set; when we integrate a term supported in K_p^j, we obtain the sum of a function of σ_j supported in $K_p + \mathbb{R}\alpha_j$ and of a function supported in K_p^j. When we integrate $t^q w(\sigma_j)$, we obtain $\frac{t^{q+1}}{q+1} w(\sigma_j)$. This completes the proof. $\quad\diamondsuit$

It turns out that the terms $v_{j0}^{(p)}(\sigma_j)$ play a special role in the approximation of u.

Definition 1.1. *The functions* $v_{j0}^{(p)}(\sigma_j)$, $j = 1, \ldots, N$, *are called the free profiles (of order (p)).*

Note that the free profiles are determined by the initial value u_0 and can be recursively computed by calculating finite integrals.

Remark finally that, for $x \ge b + \lambda_{N-1}(0)t$, all the coefficients $v_{jq}^{(p)}$ ($j \le N - 1$) in (1.4) are zero, thus u_j ($j \le N - 1$) is formally zero. The same thing happens to the formal components u_j ($j \ge 2$) in the domain $x \le a + \lambda_2(0)t$. Indeed, it is a well known fact that, next to a constant state, the solution u is a simple wave (see [CFr], [Sm] or [Hö2]). Hence the components of u are actually, and not just formally, zero.

1.2. Slow time and reduced equations

a. Construction of the approximate solution

If we collect the terms $u_j^{(p)}$ described in Theorem 1.1, we obtain

$$(1.6) \qquad u_j(x,t) = \varepsilon \sum_{0 \le q \le p-1} \varepsilon^{p-q-1}(\varepsilon t)^q v_{jq}^{(p)}(\sigma_j(x,t)) + R_j(x,t)$$

where

(1.7)
$$R_j(x,t) = \sum_{p \geq 1} \varepsilon^p r_j^{(p)}(x,t).$$

This suggests to introduce the **slow time** $\tau = \varepsilon t$ as a new variable, and to consider u_j, neglecting R_j, as a function U_j of σ_j and τ (depending of course smoothly on ε).

Two facts motivate this approach:

(i) Burgers' equation

$$\partial_t u + (\lambda + u)\partial_x u = 0, \quad u(x,0) = \varepsilon\, v_0(x)$$

has the exact solution

$$u(x,t) = \varepsilon\, v(x - \lambda t, \varepsilon t),$$

where v is the solution of

$$\partial_\tau v + v\partial_\sigma v = 0, \quad v(\sigma,0) = v_0(\sigma).$$

Thus the appearance of τ is not a big surprise.

(ii) The asymptotic expansion (1.6) does not give us any more information when t has the order of magnitude of ε^{-1}. More precisely, suppose that we have recursively computed a finite number of terms $u_j^{(p)}$, $p \leq s$; this gives us at most the terms τ^q, $q \leq s - 1$, in the Taylor expansion at $\tau = 0$ of U_j. This information is irrelevant for finite (not small) values of τ.

We argue as follows. Fix $s \in \mathbb{N}$ and set

$$u_j^s = \sum_{1 \leq p \leq s} \varepsilon^p u_j^{(p)}.$$

As soon as $t \geq T_s$, the terms $r_j^{(p)}$ in the expressions of $u_j^{(p)}$ vanish; as soon as the strips $K_s + \mathbb{R}\alpha_j$ are disjoint (say, $t > C_s$), the various components u_j^s of u^s have disjoint supports. Thus, to continue u^s for large time, we can try the **ansatz**

(1.8)
$$u_j(x,t) = \varepsilon\, w_j(\sigma_j(x,t), \tau), \quad j = 1, \ldots, N,$$

where moreover the u_j have disjoint supports.

Inserting in (1.1) and taking into account the normalization, we obtain

$$(1.9) \qquad \partial_\tau w_j + \tilde{\lambda}_j(w_j, \varepsilon)\partial_\sigma w_j = 0, \quad j = 1, \ldots, N,$$

where

$$(1.10) \qquad \tilde{\lambda}_j(w_j, \varepsilon) = \varepsilon^{-1}(\lambda_j(0, \ldots, 0, \varepsilon w_j, 0, \ldots, 0) - \lambda_j(0)).$$

We call these equations **"reduced equations"**.

What we have gained is that the system (1.1) reduces to a collection of scalar equations, which are well understood. Note that the main term in $\tilde{\lambda}_j$ (that is, the one corresponding to $\varepsilon = 0$), is $w_j\partial_j\lambda_j(0)$.

To understand what are the correct initial values on $\{\tau = 0\}$ for (1.9), we observe the two following facts:

(i) The formal solution $U_j = \varepsilon\Sigma\varepsilon^{p-q-1}\tau^q v_{jq}^p(\sigma_j)$, considered as a function of σ and τ, satisfies (1.9).

(ii) A solution of (1.9) is determined by its trace on $\tau = 0$.
The trace of U_j on $\tau = 0$ being

$$U_j(\sigma, 0) = \varepsilon \sum_{p\geq 1} \varepsilon^{p-1} v_{j0}^{(p)}(\sigma),$$

we should impose, in addition to (1.9),

$$(1.11) \qquad w_j(\sigma, 0) = \sum_{p\geq 1} \varepsilon^{p-1} v_{j0}^{(p)}(\sigma).$$

We see that the free profiles only appear in this initial condition. In other words, the equations (1.9) contain in their structure the mechanism by which the terms generated in $u_j^{(p)}$ by resonant interactions are produced.

The approximate solution is then constructed by gluing together the short time approximation u^s and the long time approximation given by (1.8):

(i) We fix s as before and set

$$u_j^s = \sum_{1\leq p\leq s} \varepsilon^p u_j^{(p)}.$$

Then we solve (1.9) and (1.11) where the sum in the right hand side is limited to $1 \leq p \leq s$, and we denote the corresponding solutions by w_j^s.

(ii) With $\chi \in C^\infty(\mathbb{R})$, $\chi(\eta) = 1$ for $\eta \leq 0$ and $\chi(\eta) = 0$ for $\eta \geq 1$, we set

(1.12) $\quad \tilde{u}_j^s(x,t) = \chi(t - C_s)u_j^s(x,t) + (1 - \chi(t - C_s))\varepsilon w_j^s(\sigma_j(x,t), \tau)$.

The function $\tilde{u}^s$ is our approximate solution.

b. lifespan of the approximate solution

Proposition 1.2. *Assume that all the eigenvalues of the system are genuinely nonlinear (that is, $\partial_j \lambda_j(0) \neq 0$). Then the approximate solution $\tilde{u}^s$ is defined for $0 \leq t < \tilde{T}_\varepsilon^s$, with*

(1.13) $\qquad \varepsilon \tilde{T}_\varepsilon^s = \inf M_j + O(\varepsilon), \quad M_j^{-1} = \max -\partial_j \lambda_j(0)(u_0^{(1)})_j'$.

PROOF OF PROPOSITION 1.2. For each equation (1.9), the results from Chapter IV (1. a) show that the solution w_j has a lifespan of inverse

$$\max -\partial_j \lambda_j(0) \partial_\sigma w_j(\sigma, 0) + O(\varepsilon).$$

Moreover,

$$w_j(\sigma, 0) = \sum_{1 \leq p \leq s} \varepsilon^{p-1} v_{j0}^{(p)}(\sigma)$$

and

$$v_{j0}^{(1)} = (u_0^{(1)})_j.$$

Hence we obtain (1.13), since $\tau = \varepsilon t$. $\qquad\qquad \Diamond$

Remark that in the linearly degenerate case (that is, $r_j(u)\nabla\lambda_j(u) \equiv 0$), we have simply $\tilde{\lambda}_j \equiv 0$.

1.3. Existence, approximation and blowup

We will state here without details some results which can be obtained, starting from the above construction of the approximate solution $\tilde{u}^s$ (see [A19]).

We distinguish here two steps:

(i) An "existence and approximation" step, where information about the true solution for large time can be obtained from the knowledge of the approximate solution. In this step however, we still stay away from the actual blowup region.

(ii) A "representation" step, where we solve the blowup system (in the sense of Chapter I) in a strip $\{\tau_1 \le \varepsilon t \le \tau_2\}$ containing the actual blowup time.

The facts of interest concerning the lifespan and the mechanism of the blowup of u are then obtained in a standard fashion as in Chapter I.

a. Existence and approximation

Though a lower bound

$$\bar{T}_\varepsilon \ge C\varepsilon^{-1}, \quad C > 0$$

is easily obtained for any system, Proposition 1.2 and the structure of $\tilde{u}^s$ make it possible to show

(1.14) $\liminf \varepsilon \bar{T}_\varepsilon \ge \inf M_j,$

where the M_j are defined in (1.13).

Moreover, we obtain for any $A < \inf M_J$ and $0 < \varepsilon \le \varepsilon_A$ a qualitative description of u.

The strategy of the proof is as follows:

(i) We first note that

$$\tilde{u}^s(x,0) = u_0(x) + O(\varepsilon^{s+1}).$$

(ii) We then have to check that

$$\partial_t \tilde{u}^s + A(\tilde{u}^s)\partial_x \tilde{u}^s = \tilde{f}^s$$

is indeed small. This is clear when $t \le C_s$, where $\tilde{f}^s = O(\varepsilon^{s+1})$, and when $t \ge C_s + 1$, where $\tilde{f}^s = 0$. In the transition period $C_s \le t \le C_s + 1$, it is sufficient to verify that u_j^s is close to εw_j^s, as a consequence of (1.11).

(iii) Finally, setting $u = \tilde{u}^s + \dot{u}$, we must show that $\dot{u}$ exists and is small. Since $\dot{u}$ satisfies a system close to the linearized system on $\tilde{u}^s$, it is not possible to keep control of this system when approaching too close the lifespan $\tilde{T}_\varepsilon^s$ of $\tilde{u}^s$. However, for any $A < \inf M_j$ and ε small enough, no blowup of w_j occurs in (1.9), and one obtains existence for $0 \le t \le A\varepsilon^{-1}$, thus proving (1.14). This argument gives also the smallness of $\dot{u}$, in the form

$$\partial_{x,t}^{\alpha}\dot{u} = O(\varepsilon^{s-N_0}),$$

for some N_0 fixed.

Again, the large time existence of $\dot{u}$ and its smallness are obtained here by induction on time (as in the proof of Theorem 4.1 of Chapter IV). It is important to remark that these two aspects are connected, the existence proof becoming easy precisely because of the smallness of $\dot{u}$.

Note also that this argument is **independent** of (and much rougher than) the estimates of Theorem 4.1. This is why it can be used also in multidimensional situation (see section 2).

b. Representation and Blowup

For simplicity, we explain this step only for $N = 3$.

Let Γ_a^2 and Γ_b^2 be the two-characteristics of the system issued from the points $(a,0)$ and $(b,0)$. As already mentioned in 1.1, the solution u is a simple wave right of Γ_b^2 and left of Γ_a^2; taking into account the normalization of the system, this just means $u_1 = u_2 = 0$ right of Γ_b^2 and $u_2 = u_3 = 0$ left of Γ_a^2. Hence the system "to the right" or "to the left" reduces in fact to a scalar equation, for which blowup mechanisms are well understood. We are interested in the case of blowup "in the middle", hence we assume

$$M_2 < \inf(M_1, M_3).$$

For convenience, we also take $\lambda_2(0) = 0$.

Using the slow time variable $\tau = \varepsilon t$, the system becomes

(1.15) $\qquad \varepsilon\partial_\tau\tilde{u} + A(\tilde{u})\partial_x\tilde{u} = 0, \ \ \tilde{u}(x,\tau) = u(x,\tau\varepsilon^{-1}).$

The blowup system (for λ_2) of this system is

$$\varepsilon\partial_T\phi = \lambda_2(\tilde{v}), \ \ {}^t\ell_2(\tilde{v})\partial_T\tilde{v} = 0,$$

(1.16) $\qquad {}^t\ell_i(\tilde{v})\big[\varepsilon\partial_X\phi\,\partial_T + (\lambda_i - \lambda_2)(\tilde{v})\partial_X\big]\tilde{v} = 0, \ \ i = 1,3.$

We fix

$$0 < \tau_0 < M_2 < \tau_1.$$

and denote by $\alpha(\varepsilon)$, $\beta(\varepsilon)$ the intersection points of Γ_a^2 and Γ_b^2 with the line $\{\varepsilon t = \tau_0\}$.

The representation of the solution u is described in the following theorem.

Theorem 1.2. *Let $p \in \mathbb{N}$, $p \geq 1$.*

(a) *For ε small enough, there exists a solution $(\tilde{v}, \phi)$ of the blowup system (1.16) in the rectangle*

$$R = \left\{ (X, T), \alpha \leq X \leq \beta, \tau_0 \leq T \leq \tau_1 \right\},$$

of the form

$$\tilde{v} = \varepsilon D_\varepsilon w, \quad D_\varepsilon = \mathrm{diag}(\varepsilon^p, 1, \varepsilon^p),$$

w being a smooth function of $(X, T, \varepsilon) \in R \times [0, \varepsilon_0]$.

(b) *Define the blowup time $\tau(\varepsilon)$ by*

$$\tau(\varepsilon) = \max \left\{ \tau, \tau_0 \leq \tau \leq \tau_1, (X, T) \in R \cap \{0 \leq T < \tau\} \right.$$
$$\left. \Rightarrow \partial_X \phi(X, T) \neq 0 \right\}.$$

Let $\tilde{u}$ be the solution of (1.15) defined for $\tau < \tau(\varepsilon)$ by

$$\tilde{u}(\phi(X, T), T) = \tilde{v}(X, T),$$

and $U(x, t) = \tilde{u}(x, \varepsilon t)$. Then, for $\tau_0 \leq \varepsilon t < \tau(\varepsilon)$ and between Γ_a^2 and Γ_b^2, $U = u$.

The point of this approach is that the determination of the lifespan follows from a glance at the zeroes of $\partial_X \phi$, defined in a "big" rectangle. In particular, in nondegenerate cases, the lifespan $\bar{T}_\varepsilon = \varepsilon^{-1} \tau(\varepsilon)$ can be obtained by perturbation of the case $\varepsilon = 0$, using the implicit function theorem.

Assume more precisely that the minimum of the function $\partial_2 \lambda_2(0)(u_0^{(1)})_2'$ defining M_2 is unique and nondegenerate (that is, the second order derivative is nonzero). Then the function $\tau(\varepsilon)$ is smooth and coincides to any order in ε with the lifespan formally computed from the theory of nonlinear geometrical optics developed in Sections 1.1/1.2. Moreover, the solution u is, near its then unique blowup point, an "exterior cusp solution" in the sense of Chapter I. An important feature of this strategy "approximation/representation" is that it makes sense also in multidimensional situations, as we shall see next.

2. Quasilinear wave equations

We will now develop the same ideas as in Section 1 for wave equations in multidimensional situations.

We consider, for $(t = x_0, x) \in \mathbb{R}^{n+1}$ $(n = 2, 3)$, the quasilinear wave equation

$$(2.1) \qquad \sum_{0 \leq i,j \leq n} g_{ij}(\nabla u)\partial^2_{ij} u = 0, \quad u(x,0) = f(x, \varepsilon), \quad \partial_t u(x,0) = g(x, \varepsilon),$$

where

$$\nabla u = (\partial_0 u, \nabla_x u), \quad g_{ij} = g_{ji}, \quad g_{00} \equiv 1,$$
$$i \geq 1 \Rightarrow g_{i0}(0) = 0, \quad i, j \geq 1 \Rightarrow g_{ij}(0) = -\delta_{ij}.$$

As usual, we suppose f, g smooth, supported in $|x| \leq M$ and satisfying

$$f(x, \varepsilon) = \varepsilon f^{(1)}(x) + \varepsilon^2 f^{(2)}(x) + \ldots, g(x, \varepsilon) = \varepsilon g^{(1)}(x) + \varepsilon^2 g^{(2)}(x) + \ldots.$$

The function u being small, the equation is thus a perturbation of the wave equation, and we set

$$g_{ij}(\eta) = g_{ij}(0) + \Sigma g^k_{ij}\eta_k + O(\eta^2).$$

We restrict our attention to $n \leq 3$, since for $n \geq 4$, Klainerman has shown the existence of a global smooth solution of (2.1).

2.1. Formal analysis

We will look for u in the form

$$u = \sum_{p \geq 1} \varepsilon^p u^{(p)},$$

where the functions $u^{(p)}$ satisfy the equations

$$(2.2)_1 \qquad (\partial^2_t - \Delta_x)u^{(1)} = 0,$$
$$(2.2)_p \qquad (\partial^2_t - \Delta_x)u^{(p)} = Q^{(p)}, \quad p \geq 2,$$

$Q^{(p)}$ being a polynomial expression of the $u^{(q)}$ $(q \leq p - 1)$. In particular,

$$(2.3) \qquad Q^{(2)} = -\Sigma g^k_{ij} \partial_k u^{(1)} \partial^2_{ij} u^{(1)}.$$

Our aim is to describe, as in Theorem 1.1, the structure of all the $u^{(p)}$. This can be done completely in a domain

$$|x| \geq t + M - C,$$

for any fixed C. Here, we will discuss only $u^{(1)}$ and $u^{(2)}$. The following classical result describes $u^{(1)}$.

Lemma 2.1. *The solution v of the wave equation with smooth initial data f, g supported in $|x| \leq M$ can be written, for $|x| \geq 2M$, in the form*

$$(2.4) \qquad u(x,t) = r^{-\frac{1}{2}(n-1)} F(r - t, \omega, r^{-1}).$$

Here $r = |x|$, $x = r\omega$ and F is a smooth function satisfying

$$(2.5) \qquad |\partial_\rho^\alpha \partial_\omega^\beta \partial_z^\gamma F(\rho, \omega, z)| \leq C_{\alpha\beta\gamma} (1 + |\rho|)^{-\frac{1}{2}(n-1) - |\alpha| + |\gamma|}.$$

Moreover,
$$(2.6)$$
$$F_0(\rho, \omega) = F(\rho, \omega, 0) = \frac{1}{2} (2\pi)^{-\frac{1}{2}(n-1)} \chi_-^{-\frac{1}{2}(n-1)} * [R(s, \omega, g) - \partial_s R(s, \omega, f)].$$

Here the star denotes convolution in s, $\chi_-^\lambda = \frac{x_-^\lambda}{\Gamma(\lambda+1)}$, $x_-^\lambda = |x|^\lambda \mathbf{1}_{x<0}$ and

$$R(s, \omega, f) = \int_{x\omega = s} f(x) \, dS(x)$$

is the Radon transform of f.

All we need to know here is

$$(2.7) \qquad u^{(1)}(x,t) \sim r^{-\frac{1}{2}(n-1)} F_0(r - t, \omega), \quad r \to +\infty, \quad r - t \geq M - C,$$

the free profile F_0 corresponding to the initial data $f^{(1)}$, $g^{(1)}$ according to lemma 2.1.

The following lemma gives a partial description of $u^{(2)}$.

Lemma 2.2. *Define, with $\omega_0 = -1$,*

$$(2.8) \qquad g(\omega) = \sum_{0 \leq i,j,k \leq n} g_{ij}^k \omega_i \omega_j \omega_k.$$

and assume

$$(ND) \qquad g(\omega) \not\equiv 0.$$

Then for $|x| \geq t + M - C$ and $n = 2$,

(2.9) $$u^{(2)}(x,t) = \frac{g(\omega)}{2}(\partial_\rho F_0)^2 (r - t, \omega) + O(r^{-\frac{1}{2}}).$$

For $n = 3$,

(2.10) $$u^{(2)}(x,t) = \frac{g(\omega)}{4r} \ell nt (\partial_\rho F_0)^2 (r - t, \omega) + O(r^{-1}).$$

SKETCH OF THE PROOF OF LEMMA 2.2.

a. From the asymptotic properties of $u^{(1)}$ described in Lemma 2.1, we obtain

$$Q^{(2)}(x,t) = -g(\omega) r^{-(n-1)} (\partial_\rho F_0 \, \partial_\rho^2 F_0)(r - t, \omega) + O(r^{-n}).$$

Note that in the domain we consider,

$$t^\lambda - r^\lambda = O(r^{\lambda - 1}).$$

b. For $n = 2$, a straightforward computation gives

$$(\partial_t^2 - \Delta)\left\{ \frac{g(\omega)}{2}\left(\frac{t}{r}\right)^{\frac{1}{2}} (\partial_\rho F_0)^2 (r - t, \omega)\right\} = Q^{(2)}(x,t) + O(r^{-2}),$$

because any function of the form

$$r^{-\frac{1}{2}} H(r - t, \omega)$$

is an approximate solution of the wave equation.

c. For $n = 3$, we have similarly

$$(\partial_t^2 - \Delta)\left\{ \frac{g(\omega)}{4r} \ell nt (\partial_\rho F_0)^2 (r - t, \omega)\right\} = Q^{(2)}(x,t) + O(r^{-3} \ell nr).$$

d. The proof can be completed by showing that the differences between $u^{(2)}$ and the main terms displayed in (b), (c) decay at least as free solutions do. $\diamond$

2.2. Slow time and reduced equation

The rough idea of the structure of the terms $u^{(p)}$ given in the Lemmas 2.1 and 2.2 is sufficient to guess the slow time of the problem and compute the reduced equations.

a. Case $n = 3$

In this case, the strong Huygens' principle holds, that is, free solutions with initial data supported in $|x| \leq M$ vanish for $|x| \leq t - M$. Thus we do not have to worry about the estimates of Lemma 2.1 and 2.2, valid only for $|x| \geq t + M - C$. In particular, F_0 is supported in $|\rho| \leq M$.

The approximation

$$u(x, t) = \varepsilon r^{-1} \left[F_0(r - t, \omega) + \varepsilon \ell n t \frac{g(\omega)}{4} (\partial_\rho F_0)^2 (r - t, \omega) + \ldots \right]$$

suggests the **slow time** $\tau = \varepsilon \ell n t$ and the ansatz

$$(2.11) \qquad u(x, t) = \varepsilon r^{-1} w(r - t, \omega, \tau).$$

Inserting (2.11) into (2.1) gives

$$(2.12) \qquad \Sigma g_{ij}(\nabla u) \partial_{ij}^2 u = \frac{\varepsilon^2}{rt} \left\{ - 2\partial_{\rho\tau}^2 w + g(\omega)(\partial_\rho w)(\partial_\rho^2 w) + \ldots \right\},$$

where the dots stand for some smaller quadratic expression of second order derivatives of w.

Thus our approximate solution is

$$(2.13) \qquad \tilde{u}(x, t) = \varepsilon \left[\chi(\varepsilon t) u^{(1)}(x, t) + (1 - \chi(\varepsilon t)) r^{-1} w(r - t, \omega, \tau) \right],$$

where

$$\chi \in C^\infty(\mathbb{R}), \quad s \leq 1 \Rightarrow \chi(s) = 1, \quad s \geq 2 \Rightarrow \chi(s) = 0$$

and w is the solution of

$$(2.14) \qquad \partial_\tau w - \frac{g(\omega)}{4} (\partial_\rho w)^2 = 0, \quad w(\rho, \omega, 0) = F_0(\rho, \omega).$$

In (2.14), ω is just a parameter, so that the results of Chapter IV, 1 show that w, hence $\tilde{u}$ exist for

$$(2.15) \qquad 0 \leq t < \exp \varepsilon^{-1} \bar{\tau}, \quad (\bar{\tau})^{-1} = \max \frac{1}{2} \Sigma g_{ij}^k \omega_i \omega_j \omega_k \partial_\rho^2 F_0(\rho, \omega).$$

b. Case $n = 2$

The approximation

$$u(x, t) = \varepsilon r^{-\frac{1}{2}} \left[F_0(r - t, \omega) + \varepsilon t^{\frac{1}{2}} \frac{g(\omega)}{2} (\partial_\rho F_0)^2 (r - t, \omega) + \ldots \right]$$

suggests the **slow time** $\tau = \varepsilon t^{\frac{1}{2}}$ and the ansatz

$$(2.16) \qquad u(x,t) = \varepsilon r^{-\frac{1}{2}} w(r - t, \omega, \tau).$$

Lemma 2.3. *Inserting (2.16) into (2.1) gives*

$$(2.17) \quad \Sigma g_{ij}(\nabla u)\partial^2_{ij}u = \frac{\varepsilon^2}{(rt)^{\frac{1}{2}}}\left\{ -\partial^2_{\rho\tau}w + g(\omega)(\partial_\rho w)(\partial^2_\rho w) + \varepsilon^2 R(w) \right\},$$

where $R(w)$ is a quadratic expression, with smooth coefficients bounded for $0 < \tau_0 \leq \tau \leq \tau_1$, of derivatives $\partial^\alpha_{\rho,\omega,\tau} w$ ($|\alpha| \leq 2$).

In this case, the strong Huygens' principle does not hold, but free solutions are still smaller far inside the light cone than they are in the domain $|x| \geq t + M - C$ (because of (2.5)). Thus we take for our approximate solution

$$(2.18)$$
$$\tilde{u}(x,t) = \varepsilon \left[\chi(\varepsilon t)u^{(1)}(x,t) + (1 - \chi(\varepsilon t))\chi(3\varepsilon(t - r))r^{-\frac{1}{2}} w(r - t, \omega, \tau) \right],$$

where the cutoff χ is as above and w is the solution of

$$(2.19) \qquad \partial_\tau w - \frac{g}{2}(\partial_\rho w)^2 = 0, \quad w(\rho, \omega, 0) = F_0(\rho, \omega).$$

As in (a), we see that $\tilde{u}$ exists for

$$(2.20) \qquad 0 \leq t < \varepsilon^{-2}\bar{\tau}^2, \quad (\bar{\tau})^{-1} = \max \Sigma g^k_{ij}\omega_i\omega_j\omega_k\partial^2_\rho F_0(\rho, \omega).$$

2.3. Existence and approximation, null conditions, blowup

As in Section 1.3, but now for the wave equation, we will indicate some consequences of the above constructions.

a. Existence and approximation

With exactly the same strategy as indicated in 1.3, one can prove

$$(2.21) \qquad \qquad \liminf \varepsilon \ell n \bar{T}_\varepsilon \geq \bar{\tau},$$

where $\bar{\tau}$ is given by (2.15) (in space dimension two,

$$(2.22) \qquad \qquad \liminf \varepsilon^2 \bar{T}_\varepsilon \geq \bar{\tau}^2,$$

where $\bar{\tau}$ is given by (2.20)).

Here of course, to control the rest $\dot{u} = u - \tilde{u}$, one has to use energy inequalities instead of the pointwise estimates available in one space dimension. The main difference lies in the fact that the solution decays as $t \to +\infty$ (a feature responsible for the large existence time); to recover these decay properties from energy inequalities, one has to use the technique of "Klainerman fields" (see [Kl2]).

b. Null conditions

If $g(\omega) \equiv 0$, we say that the null condition is satisfied. In this case, the main terms of $u^{(2)}$ in (2.9), (2.10) vanish, and the reduced equations (2.14), (2.19) become linear. This suggests that the lifespan should be much bigger under these circumstances. This situation is analogous to the situation of a first order system with linearly degenerate eigenvalues (see 1.2).

In dimension three, Klainerman and Christodoulou have proved global existence when the null condition is satisfied. In dimension two, the situation is more complicated:

(i) If all the g_{ij}^k vanish, Hoshiga has proved

$$\liminf \varepsilon^2 \ln \bar{T}_\varepsilon \geq \bar{\zeta},$$

where the number $\bar{\zeta}$ can be explicitly computed from the second order derivatives of g_{ij} and the free profile F_0 (defined in (2.6)). This result is similar to (2.21), (2.22), with a slow time

$$\zeta = \varepsilon^2 \ln t.$$

We will come back to this in Section 3.

(ii) If moreover $\bar{\zeta} = \infty$ (which corresponds to a second null condition), global existence holds.

The general case $g(\omega) \equiv 0$ does not seem to have been settled yet. Note that in all cases, including when global existence holds, the approximate solution provides useful information.

c. Blowup

In the special case of rotationally invariant wave equations handled in Chapter IV, we obtained (Theorem 5.1, (5.5)) the upper bound

$$\limsup \varepsilon \, \ell \, n \, \bar{T}_\varepsilon \le \bar{\tau}, \quad (\bar{\tau})^{-1} = \max c'(0) \, F''(\rho).$$

In this case, $g = 2c'(0)$, hence this upper bound agrees with the one suggested by (2.15) (and similarly in dimension two).

Moreover, the blowup mechanism suggested by (2.13), (2.14) is

$$|\nabla^2 u| \to +\infty.$$

This agrees with what we have learned of first order systems in Chapter IV; in this case, only the gradient of the solution is expected to blowup. If we reduced the scalar equation on u to a first order system on U, $\nabla^2 u$ would correspond to ∇U.

When using the strategy explained in 1.3., one runs into the following difficulty: we do not know how to solve the blowup equation for a general class of initial data. Thus, in sharp contrast with the one dimensional case, it is still an open problem to find a relevant upper bound to $\bar{T}_\varepsilon$ and, a fortiori, to describe the blowup mechanism.

In the next section, we will sketch some asymptotic approach to this problem, where the blowup equation is only asymptotically (and not exactly) solved.

3. Further results on the wave equation

For the case of one space dimension handled in Section 1. (Theorem 1.1), we were able to compute explicitly all the terms of a formal asymptotic solution. For the quasilinear wave equations of Section 2, we computed only approximations of the first two terms in a zone $\{|x| \ge t + M - C\}$. In particular, no other free profile than F_0 (defined in (2.6)) was introduced.

We want to sketch here how the program of Section 1.3 (formal analysis, slow time and reduced equations, approximation for large time and representation of the solution as a blowup solution) can be carried out, and how it makes it possible to reduce the problem of blowup for (2.1) to a **local blowup problem** (3.4). We discuss a model case of this problem in Section 3.3, and return finally to our original problem in Section 3.4, where our blowup results of asymptotic nature are stated.

For simplicity, we restrict our attention to the case $n = 2$.

3.1. Formal analysis near the boundary of the light cone

We use again the notations of 2.1. It turns out to be difficult to describe the functions $u^{(p)}$ inside the light cone $\{|x| = t\}$, especially in the case $n = 2$ where the strong Huygens's principle does not hold. We skip this difficulty and complete the partial information of Lemma 2.2.

Proposition 3.1. *For all $C > 0$, $N, N' \in \mathbb{N}$, there exist functions*

$$L_q^{(p)}(\rho, \omega, z), \ R_{q,q'}^{(p)}(\rho, \omega, z), \ \rho = r - t, \ z = r^{-1}$$

such that the term $u^{(p)}$ can be written

$$u^{(p)}(x, t) = r^{-\frac{1}{2}} \Big\{ \sum_{0 \leq 2q \leq p-1} (\ell n\, t)^q L_q^{(p)}(r - t, \omega, z) +$$

(3.1)
$$+ \sum_{q \geq 1, q+2q' \leq p-1} t^{\frac{q}{2}} (\ell n\, t)^{q'} R_{q,q'}^{(p)}(r - t, \omega, z) \Big\} + r^{(p)}.$$

Here, for $|x| \geq t + M - C, t \geq C$,

(i) $r^{(p)} = O(r^{-N})$,

(ii) $(\partial_t^2 - \Delta) r^{-\frac{1}{2}} L_q^{(p)} = O(r^{-N'})$,

(iii) For $q = 2k$ $(k \geq 1)$, the term $R_{q,0}^{(p)}(\rho, \omega, z)$ does not contain the power z^k.

For instance,

$$u^{(1)}(x, t) = r^{-\frac{1}{2}} L_0^{(1)}(r - t, \omega, z) + r^{(1)},$$

which is just (2.4); also

$$u^{(2)}(x, t) = r^{-\frac{1}{2}} \Big\{ L_0^{(2)}(r - t, \omega, z) + t^{\frac{1}{2}} R_{1,0}^{(2)}(r - t, \omega, z) \Big\} + r^{(2)},$$

which completes (2.9).

Just as in the one dimensional case (Definition 1.1), we define the function

$$L_0^{(p)}(r - t, \omega, z)$$

as the **free profile** of order p.

3.2. Slow time and reduced equations

When we collect the terms $u^{(p)}$, we find formally

$$u(x,t) = \varepsilon\, r^{-\frac{1}{2}} \left\{ \Sigma \varepsilon^{p-1-2q}\, \zeta^q\, L_q^{(p)}(\rho,\omega,z) + \Sigma \varepsilon^{p-1-q-2q'}\, \tau^{\frac{q}{2}}\, \zeta^{q'}\, R_{q,q'}^{(p)} \right\} + \cdots .$$

Recall that

$$\rho = r - t, \quad z = r^{-1}.$$

We see the presence of **two slow times**

$$\tau = \varepsilon\, t^{\frac{1}{2}}, \quad \zeta = \varepsilon^2 \ell nt,$$

and this suggests the ansatz

$$(3.2) \qquad u(x,t) = \varepsilon\, r^{-\frac{1}{2}} F(r-t, \omega, z, \tau, \zeta).$$

As before, the equation obtained by inserting (3.2) into (2.1) has to be supplemented with some boundary conditions on F analogous to (2.19). It turns out that F is determined by its values

$$\partial_\tau^{2q}\, \partial_z^q\, F(\rho, \omega, 0, 0, 0), \quad q = 1, 2, \ldots .$$

The choice ensuring the matching of the formal asymptotics with the ansatz is

$$F(\rho, \omega, 0, 0, 0) = \Sigma \varepsilon^{(p)}\, L_0^{(p)}(\rho, \omega, 0),$$

and zero for the derivatives with $q \geq 1$.

Remark that the presence of the second slow time ζ is coherent with the results mentioned in section 2.3. b.

The main consequence of the existence of F, depending smoothly on ε, representing u as in (3.2) is the following: choose

$$0 < \tau_0 < \tau_2 < \bar{\tau} < \tau_1$$

where $\bar{\tau}$ is defined in (2.20) with the property (2.22). After some transition period, for $\tau \geq \tau_0$, u can be represented for $|x| \geq t + M - C$ in the form

$$(3.3) \qquad u(x,t) = \varepsilon\, r^{-\frac{1}{2}} w(r-t, \omega, \tau).$$

Here w denotes the old F considered now as a function of ρ, ω and τ, depending smoothly on ε and $\varepsilon^2 \ell n\varepsilon$.

This is a consequence of the identities

$$z = (\varepsilon\tau^{-1})^2(1 + \rho(\varepsilon\tau^{-1})^2)^{-1}, \quad \zeta = 2\varepsilon^2\ell n\tau - 2\varepsilon^2\ell n\varepsilon.$$

We are back now to the situation of Lemma 2.3, with the difference that we cannot neglect the term $R(w)$ in (2.17) anymore. We have to solve the following local blowup problem:

Find, for $\tau \geq \tau_0$ (and up to the blowup time), a solution w of

$$(3.4) \qquad -\partial^2_{\rho\tau}w + g(\omega)(\partial_\rho w)(\partial^2_\rho w) + \varepsilon^2 R(w) = 0$$

such that w agrees in some strip $\tau_0 \leq \tau \leq \tau_2$ with the solution of (3.4) obtained from F.

This is a purely local problem (in the slow time variable τ).

3.3. A local blowup problem

We consider now (3.4): this is a second order equation in τ, and we expect its solution w to vanish for $\rho \geq M$ (this corresponds to the zone $|x| \geq t + M$ where the solution u vanishes).

We skip the difficulty that w has to verify two trace conditions on $\{\tau = \tau_0\}$ while (3.4) is of first order in τ for $\varepsilon = 0$.

The term $R(w)$ in (3.4) corresponding to the actual wave equation (2.1) is rather complicated; but what really creates a difficulty in solving (3.4) are the $\omega-$derivatives. Thus we will concentrate now on the following **simplified model problem**: given $v_0(\rho, \omega)$, vanishing for $\rho \geq M$ and rapidly decaying for $\rho \to -\infty$, find $v(\rho, \omega, \tau)$, also vanishing for $\rho \geq M$, satisfying

$$(3.5) \qquad \partial_\tau v + v\partial_\rho v + \varepsilon\partial^2_\omega \int_M^\rho v(s, \omega, \tau)\,ds = 0, \quad v(\rho, \omega, 0) = v_0(\rho, \omega).$$

In particular, we want to obtain an asymptotic expansion of the lifespan $\bar{\tau}_\varepsilon$ of v.

We believe that this problem contains all the essential difficulties of (3.4).

a. Let us first insist on the following point. It would seem natural, to solve (3.5), to try the ansatz

$$v = v^{(0)} + \varepsilon v^{(1)} + \dots,$$

where the $v^{(j)}$ satisfy the equations

$$\partial_\tau v^{(0)} + v^{(0)} \partial_\rho v^{(0)} = 0,$$

$$\partial_\tau v^{(1)} + v^{(0)} \partial_\rho v^{(1)} + v^{(1)} \partial_\rho v^{(0)} = -\partial_\omega^2 \int_M^\rho v^{(0)} ds,$$

and so on, with the corresponding terms of v_0 for initial data. If $\bar{\tau}$ denotes the lifespan of $v^{(0)}$ (given by the formula (1.3) of Chapter IV), $\bar{\tau}$ is also the lifespan of all the $v^{(p)}$; moreover, the successive terms do not blow up uniformly for $\tau = \bar{\tau}$ as some negative power of $\bar{\tau} - \tau$, which makes it impossible to introduce an appropriate slow time as before. Thus this method yields only a poor lower bound for the lifespan.

b. A better result is obtained by computing the blowup equation corresponding to the formal change

$$\rho = \phi(X, \omega, T), \quad \omega = \omega, \quad \tau = T, \quad \phi(X, \omega, 0) = X$$

with

$$v(\rho, \omega, \tau) = V(X, \omega, T).$$

This is exactly what we have done in Chapter I, B 2, formula (2.10) and (2.11) (with a instead of ε and y instead of ω). Formally inserting

$$\phi = \phi^{(0)} + \varepsilon \phi^{(1)} + \dots, \quad V = V^{(0)} + \varepsilon V^{(1)} + \dots$$

into (2.11), we obtain a system of **linear** equations

$$\partial_T \phi^{(0)} - V^{(0)} = 0, \qquad \partial_{XT}^2 V^{(0)} = 0,$$
$$\partial_T \phi^{(p)} - V^{(p)} = E^{(p)}, \qquad \partial_{XT}^2 V^{(p)} = F^{(p)},$$

where $E^{(p)}$ and $F^{(p)}$ are differential expressions involving only $\phi^{(q)}, V^{(q)}$ for $0 \leq q \leq p - 1$.

The important point is that, just as we have done in 1.3b, we can solve these equations **globally** (that is, in the strip $\tau_0 \leq \tau \leq \tau_1$ including the approximate blowup time $\bar{\tau}$) with appropriate boundary conditions. It remains then only to look for the set

$$\{(X, \omega, T), \partial_X \phi(X, \omega, T) = 0\},$$

recalling that the lifespan is the minimum of T on this set.

Since the first term $\phi^{(0)}, V^{(0)}$ (corresponding to the solution of Burgers' equation) can be explicitly computed, the implicit function theorem will allow us to look for the zeroes of $\partial_X \phi$ close to the zeroes of $\partial_X \phi^{(0)}$, and to obtain an approximation of the lifespan to any order in ε.

When we can solve exactly (2.11), we obtain an exact representation of the solution u, thus proving blowup (and, in fact, geometric blowup). This is the case in particular if the initial value v_0 is analytic in ω.

3.4. Asymptotic lifespan for the two dimensional wave equation

In Sections 3.2 and 3.3, we have explained a strategy to prove blowup for the quasilinear wave equation based on an exact representation of the solution by means of a (coordinate) blowup. This strategy is the same as the one used in 1.3 for the one dimensional case.

Unfortunately, it is not known how to solve exactly the blowup system corresponding to (3.4) in general, thus no proof of actual blowup has yet been obtained. However, some results of **asymptotic** nature are available, corresponding to asymptotic solutions of this blowup system; these results display the "blowup behavior" of the solution.

Theorem 3.4. *Assume that the function* $-g(\omega)\partial_\rho^2 F_0(\rho, \omega)$ *has a unique negative minimum at* (ρ_0, ω_0), *with positive definite Hessian. Then there exists a function* $\bar{T}_\varepsilon^a$ *with the two following properties:*

(i) *For all* N, $\bar{T}_\varepsilon \geq \bar{T}_\varepsilon^a - \varepsilon^N$ *for* $0 < \varepsilon \leq \varepsilon_N$.

(ii) *There exists* $C > 0$ *such that, for* $(C\varepsilon)^{-2} \leq t \leq \bar{T}_\varepsilon^a - \varepsilon^N$ *and* $0 < \varepsilon \leq \varepsilon_N$,

$$(C(\bar{T}_\varepsilon^a - t))^{-1} \leq \|\nabla^2 u(\cdot, t)\|_{L^\infty} \leq C(\bar{T}_\varepsilon^a - t)^{-1}.$$

Moreover, the function $\bar{T}_\varepsilon^a$ *is of the form*

$$\bar{T}_\varepsilon^a = \varepsilon^{-2}\bar{\tau}^a(\varepsilon, \varepsilon^2 \ln \varepsilon),$$

where $\bar{\tau}^a$ *is a* C^∞ *function of its arguments with* $\bar{\tau}^a(0,0) = \bar{\tau}$ ($\bar{\tau}$ *is defined in (2.20)).*

We have in particular

$$\bar{\tau}^a = \bar{\tau} + A_1\varepsilon + O(\varepsilon^2 \ln \varepsilon).$$

The second constant is

$$A_1 = -\bar{\tau}^2 \, g(\omega_0) \, \partial_\rho^2 \, L_0^{(2)}(\rho_0, \omega_0, 0),$$

where the second free profile $L_0^{(2)}$ is defined in Theorem 3.1.

We call the function $\bar{T}_\varepsilon^a$ the **"asymptotic lifespan"**. When t approaches this value, the second order derivatives of u become very large, as expected.

Whether this is a prelude to actual blowup or just the onset of some chaotic regime is an open problem.

Notes

The material for Section 1.1 and 1.2 of Section 1 is partially taken from Majda and Rosales [MaRo] and Hörmander [Hö2]. Section 1.3 is taken from the author [Al9].

Section 2 is due to Christodoulou [Ch], Klainerman [Kl1] [Kl2], John [Jo5] (for the case $n = 3$) and Hörmander [Hö2], and follows the presentation by Hörmander again.

The results of Section 3 are taken from the author's work [Al1], [Al2], [Al3]. For results in the same direction in space dimension three, see John [Jo5].

Bibliography

[Al1] ALINHAC S., *Approximation près du temps d'explosion des solutions d'équations d'ondes quasilinéaires en dimension deux,* to appear, Siam J. Math. Anal., (1994).

[Al2] ALINHAC S., *Temps de vie et comportement explosif des solutions d'équations d'ondes quasilinéaires en dimension deux, I,* to appear, Ann. Sc. ENS, (1994).

[Al3] ALINHAC S., *Temps de vie et comportement explosif des solutions d'équations d'ondes quasilinéaires en dimension deux, II,* Duke Math. J., **73** n° 3, (1994), 543-560.

[Al4] ALINHAC S., *Temps de vie et comportement explosif des solutions d'équations d'ondes quasilinéaires en dimension deux,* Séminaire d'EDP, Ecole Polytechnique, Paris, (1993).

[Al5] ALINHAC S., *Une solution approchée en grand temps des équations d'Euler compressibles axisymétriques en dimension deux,* Comm. PDE **17** (3 and 4), (1992), 447-490.

[Al6] ALINHAC S., *Temps de vie des solutions régulières des équations d'Euler compressibles axisymétriques en dimension deux,* Invent. Math. **111**, (1993), 627-670.

[Al7] ALINHAC S., *Approximation et temps de vie des solutions des équations d'Euler isentropiques en dimension deux d'espace,* Séminaire d'EDP, Ecole Polytechnique, Paris, (1991).

[Al8] ALINHAC S., *Explosion géométrique pour des systèmes quasilinéaires,* Séminaire d'EDP, Ecole Polytechnique, Paris, (1993), and article to appear, Amer. J. Math.

[Al9] ALINHAC S., *Temps de vie précisé et explosion géométrique pour des systèmes hyperboliques quasi-linéaires en dimension un d'espace,* article to appear in Ann. Scuola N. S. Pisa (1995).

[AG] ALINHAC S. and GÉRARD P., *Opérateurs pseudo-différentiels et théorème de Nash-Moser,* Inter Editions, Paris, (1991).

[Ar] ARNOLD V., *Chapitres supplémentaires de la théorie des équations différentielles ordinaires,* Editions de Moscou, (1978).

[BG] BAOUENDI M.S. and GOULAOUIC C., *Cauchy problems with characteristic initial hypersurface,* Comm. Pure Appl. Math. XXVI, (1973), 455-475.

[BKM] BEALE J.T., KATO T. and MAJDA A., *Remarks on the breakdown of smooth solutions for the 3-D Euler equations,* Comm. Math. Phys., **94**, (1984), 61-66.

[CF1] CAFARELLI L. and FRIEDMAN A., *The blow up boundary for nonlinear wave equations,* Trans. Amer. Math. Soc., **297** (1), (1986), 233-241.

[CF2] CAFARELLI L. and FRIEDMAN A., *Differentiability of the blow up curve for one dimensional nonlinear wave equations,* Arch. Rat. Mech. Anal., **91**, 1, (1985), 83-98.

[Ch] CHEMIN J.Y., *Remarques sur l'apparition de singularités fortes dans les écoulements compressibles,* Comm. Math. Phys., **133**, (1990), 323-329.

[CB] CHOQUET-BRUHAT Y., *Non existence de solutions globales de certaines équations d'onde non linéaires sur une variété,* C. R . Acad. Sci. Paris, **305** (1987), 817-821.

[Ch] CHRISTODOULOU D., *Global solutions of nonlinear hyperbolic equations for small initial data,* Comm. Pure Appl. Math. **39**, (1986), 267-282.

[CFr] COURANT R. and FRIEDRICHS K.O., *Supersonic flow and shock waves,* Springer Verlag, New York, (1949).

[DM] DI PERNA R. and MAJDA A., *The validity of nonlinear geometric optics for weak solutions of conservation laws,* Comm. Math. Phys. **98**, (1985), 313-347.

[F] FRIEDLANDER F.G., *On the radiation field of pulse solutions of the wave equation I, II,* Proc. Roy. Soc. A., **269** (1962), 53-65, and **279**, (1964), 386-394.

[GR] GÉRARD P. and RAUCH J., *Propagation de la régularité locale de solutions d'équations hyperboliques non linéaires,* Ann. Inst. Fourier **37** (3), (1987), 65-84.

[Gl1] GLASSEY R., *Blow up theorems for nonlinear wave equations,* Math. Z., **132** (1973), 183-203.

[Gl2] GLASSEY R., *Finite time blow up for solutions of nonlinear wave equations,* Math. Z., **177** (1981), 323-340.

[GL] GLIMM J. and LAX P.D., *Decay of solutions of systems of nonlinear hyperbolic conservation laws,* Memoirs Amer. Math. Soc. **101**, (1970).

[Go] GODIN P., *Lifespan of solutions of semilinear wave equations in two space dimensions,* Comm. in PDE, **18** (5 and 6), (1993), 895-916.

[GG] GOLUBITSKY M. and GUILLEMIN V., *Stable mappings and their singularities,* Graduate Texts in Math. **14**, Springer Verlag, New York, (1973).

[HJ] HANOUZET B. and JOLY J.L., *Explosion pour des problèmes hyperboliques semilinéaires avec seconds membres non compatibles,* C.R. Acad. Sci. Paris, **301** (1985), 581-584.

[Hö1] HÖRMANDER L., *Nonlinear hyperbolic differential equations,* Lectures, University of Lund, (1986-1987).

[Hö2] HÖRMANDER L, *The lifespan of classical solutions of nonlinear hyperbolic equations,* Lecture Notes Math. **1256**, Springer Verlag, (1986), 214-280.

[Ho1] HOSHIGA A., *The initial value problems for quasilinear wave equations in two space dimensions with small data,* Adv. Math. Sci. Appl., to appear.

[Ho2] HOSHIGA A., *Blow up of the radial solutions to the equations of vibrating membrane,* Preprint, (1994).

[Jo1] JOHN F., *Blow up of solutions of nonlinear wave equations in three space dimensions for small initial data,* Manuscripta Math., **28** (1979), 235-268.

[Jo2] JOHN F., *Formation of singularities in one dimensional nonlinear wave propagation,* Comm. Pure Appl. Math., **27**, (1974), 377-405.

[Jo3] JOHN F., *Blow up for quasilinear wave equations in three space dimensions,* Comm. Pure Appl. Math., **34** (1981), 29-51.

[Jo4] JOHN F., *Blow up of radial solutions of $u_{tt} = c^2(u_t)\Delta u$ in three space dimensions,* Mat. Apl. Comput. V, (1985), 3-18.

[Jo5] JOHN F., *Solutions of quasilinear wave equations with small initial data. The third phase,* Lecture Notes in Math. **1402**, Springer Verlag, (1989), 155-173.

[Jo6] JOHN F., *Nonlinear wave equations. Formation of singularities,* Lehigh University, University Lecture Series, Amer. Math. Soc., Providence, (1990).

[Ka] KATO T., *Blow up of solutions of some nonlinear hyperbolic equations,* Comm. Pure Appl. Math., **32**, (1980), 501-505.

[Ke] KELLER J., *On solutions of nonlinear wave equations,* Comm. Pure Appl. Math., **10**, (1957), 523-530.

[KL] KICHENASSAMY S. and LITTMAN W., *Blow up Surfaces for Nonlinear Wave Equations I and II,* Comm. PDE, **18** (1993), 431-452, and 1869-1899.

[Kl1] KLAINERMAN S., *The null condition and global existence to nonlinear wave equations,* Lect. Appl. Math. **23**, (1986), 293-326.

[Kl2] KLAINERMAN S., *Uniform decay estimates and the Lorentz invariance of the classical wave equation,* Comm. Pure Appl. Math. **38**, (1985), 321-332.

[KM] KLAINERMAN S. and MAJDA A., *Formation of singularities for wave equations including the nonlinear vibrating string,* Comm. Pure Appl. Math., **33**, (1980), 241-263.

[La1] LAX P.D., *Development of singularities of solutions of nonlinear hyperbolic partial differential equations,* J. Math. Physics **5** (5), (1964), 611-613.

[La2] LAX P.D., *Hyperbolic systems of conservation laws II,* Comm. Pure Appl. Math., **10**, (1957), 537-566.

[La3] LAX P.D., *Hyperbolic systems of conservation laws and the mathematical theory of shock waves,* Regional Conf. Series in Appl. Math. **13**, SIAM, (1973).

[Le] LEBAUD M.P., *Description de la formation d'un choc dans le p-système,* J. Math. Pure Appl., to appear.

[Lei] LEICHTNAM E., *Construction de solutions singulières pour des équations aux dérivées partielles non linéaires,* Ann. Sc. ENS, 4ième série, tome 20, (1987), 137-170.

[LFX] LE FLOCH P. and XIN Z., *Formation of singularities in periodic solutions to gas dynamics equations,* Preprint, (1993).

[Le1] LEVINE H., *Non existence of global weak solutions to some properly and improperly posed problems of mathematical physics: the method of unbounded Fourier coefficients,* Math. Ann., **214**, (1975), 205-220.

[Le2] LEVINE H., *The role of critical exponents in blow up theorems,* SIAM Review, **32**, (1990), 262-288.

[Lin] LINBLAD H., *Blow up of solutions of $\partial_t^2 u - \Delta u = |u|^p$ with small initial data,* Comm. in PDE, **15** (6), (1990), 757-821.

[Li] LIU T.P., *Development of singularities in the nonlinear waves for quasilinear hyperbolic partial differential equations,* J. Diff. Equ. **33**, (1979), 92-111.

[Ma] MAJDA A., *Compressible fluid flow and systems of conservation laws,* Springer Appl. Math. Sc., **53**, (1984).

[MR] MAJDA A. and ROSALES R., *Resonantly interacting weakly nonlinear waves I, a single space variable,* Stud. Appl. Math. **71**, (1984), 149-179.

[Me] MÉTIVIER G., *Problèmes de Cauchy et ondes non linéaires,* Journées Equations aux Dérivees Partielles, Saint Jean de Monts, (1986).

[Na] NATALINI R., *Unbounded solutions for conservation laws with source,* Nonlinear Anal., Th. Meth. Appl. **21** (5), (1993), 349-362.

[Rah1] RAMMAHA M., *Finite time blow up for nonlinear wave equations in high dimensions,* Comm. PDE **12** (6), (1987), 677-700.

[Rah2] RAMMAHA M., *On the blowing up of solutions to nonlinear wave equations in two space dimensions,* J. Reine Ang. Math. **391**, (1988), 55-64.

[Rah3] RAMMAHA M., *Formation of singularities in compressible fluids in two space dimensions,* Proc. Amer. Math. Soc. **107** (3), (1989), 705-714.

[Ra]   RAUCH J., *Symmetric positive systems with boundary characteristic of constant multiplicity,* Trans. Amer. Math. Soc., **291** (1), (1985), 165-187.

[Sa]   SABLÉ-TOUGERON M., *Justification de l'optique géométrique faiblement non linéaire pour le problème mixte : cas des concentrations,* preprint, (1993).

[Sc]   SCHAEFFER J., *Finite time blow up for $\partial_t^2 u - \Delta u = H(u_r, u_t)$ in two space dimensions,* Comm. PDE, **11** (5), (1986), 513-543.

[Sh1] SHATAH J., *Weak solutions and the developement of singularities in the $SU(2)$ $\sigma-model,$* Comm. Pure Appl. Math., **41**, (1988), 459-469.

[Sh2] SHATAH J. and STRUWE M., *Regularity results for nonlinear wave equation,* Annals of Math., (1993), to appear.

[Si1] SIDERIS T., *Non existence of global solutions to semilinear wave equations in high dimensions,* J. Diff. Equ., **52**, (1984), 378-406.

[Si2] SIDERIS T., *Formation of singularities in solutions to nonlinear hyperbolic equations,* Arch. Rat. Mech. Anal., **86** (4), 1984, 369-381.

[Si3] SIDERIS T., *Formation of singularities in three dimensional compressible fluids,* Comm. Math. Phys., **101**, (1985), 475-48

[Sm]   SMOLLER J., *Shock waves and reaction diffusion equations,* Grundl. **258**, Springer Verlag, New York, (1983).

[St]   STRAUSS W., *Nonlinear wave equations,* Conf. Board Math. Sc., **73**, Amer. Math. Soc., (1989).

[Wa]   WASOW W., *Asymptotic expansions for ordinary differential equations,* Krieger, New-York, (1976).

[Zu]   ZUILY C., *Solutions en grand temps d'équations d'ondes non linéaires,* Sémi- naire Bourbaki **779**, Paris, (1993/1994).

Index